Petrophysical Evaluation of Hydrocarbon Pore-Thickness in Thinly Bedded Clastic Reservoirs

By

Q. R. Passey, K. E. Dahlberg, K. B. Sullivan, H. Yin,
R. A. Brackett, Y. H. Xiao, and A. G. Guzmán-Garcia

AAPG Archie Series, No. 1

Published by

The American Association of Petroleum Geologists
Tulsa, Oklahoma, U.S.A.

ISBN: 978-0-89181-750-5
0-89181-750-6

AAPG Editor: Ernest A. Mancini
AAPG Geoscience Director: James B. Blankenship
Production: ProType Inc., Tulsa, Oklahoma; Printed in China.

On the cover: Deltaic Turbidites, Carboniferous, Central Clare Group, Tullig Cyclothem: Loophead Peninsula, Truskheve, Ireland. Photo by Rob Wellner.

This and other AAPG Publications are available from:

The AAPG Bookstore
P.O. Box 979
Tulsa, OK 74101-0979
U.S.A.
Telephone: 1-918-584-2555 or 1-800-364-AAPG (U.S.A. and Canada)
Fax: 1-918-560-2652 or 1-800-898-2274 (U.S.A. and Canada)
E-mail: bookstore@aapg.org
www.aapg.org

Geological Society Publishing House
Unit 7, Brassmill Enterprise Centre
Brassmill Lane, Bath BA13JN
U.K.
Telephone: +44-1225-445046
Fax: +44-1225-442836
E-mail: sales@geolsoc.org.uk
www.geolsoc.org.uk

Canadian Society of Petroleum Geologists
No. 160, 540 Fifth Avenue S.W.
Calgary, Alberta T2P 0M2
Canada
Telephone: 1-403-264-5610
Fax: 1-403-264-5898
E-mail: jaime.croft@cspg.org
www.cspg.org

Affiliated East-West Press Private Ltd.
G-1/16 Ansari Road, Darya Ganj
New Delhi 110 002
India
Telephone: +91-11-23279113
Fax: +91-11-23260538
E-mail: affiliat@vsnl.com

About the Authors

Quinn R. Passey holds a B.S. degree in Geology from Brigham Young University, and M.S. and Ph.D. degrees in Planetary Science and Geology from the California Institute of Technology. In 1982 he joined Exxon and has worked a variety of research and operations issues including: source rock evaluation from well logs, shaly sand analysis, and thin-bed evaluation. Current research is on formation evaluation of high-angle and horizontal wellbores. He has been an SPWLA Distinguished Lecturer and has served on the SPWLA Technology Committee. Previous life included studying the moons of Jupiter and Saturn and in discovering an earth-crossing asteroid. Quinn also advises NASA-JPL on methods to evaluate future drillwells on Mars.

Kenneth E. Dahlberg (PetroGeeks LLC) is an independent petrophysical consultant in Houston, Texas. He joined Exxon Production Research Company in 1982, where his first major research effort was the development of a well-log inversion application that was used company-wide for more than 10 years. Dahlberg has been an SPWLA Distinguished Lecturer, served on the SPWLA Technology Committee, and co-edited SPE reprint volume 39, *Petrophysics*. He completed 20 years of petrophysics research with ExxonMobil and retired at the end of 2001. Dahlberg's academic background is in mathematics, with BA and Ph.D. degrees from Rice and Princeton, respectively.

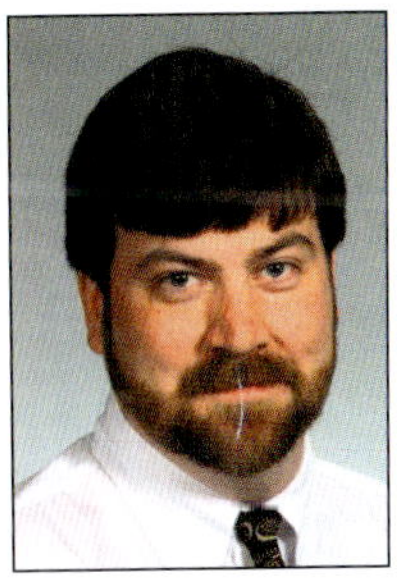

Keith Sullivan holds B.S. and M.A. degrees in Geology from the Universities of Michigan and Texas, respectively. In 1988 Sullivan joined Exxon Company, U.S.A., in Thousand Oaks, California, where he worked for 4 years as a production geoscientist. In 1992 he moved to Houston, Texas, to join a group of formation evaluation specialists in Exxon Exploration Company. In 1995 Sullivan joined Exxon Production Research Company where he contributed to a variety of petrophysical research projects relating to borehole imaging, thin-bed petrophysics and producibility prediction. In 1999 he participated in a multi-disciplinary, collaborative research project with Exxon's Upstream Development Company where he developed log-based facies models for deep-water reservoirs and log-seismic integration methodologies for 3-D geologic model construction. In 2000 Sullivan returned to Exploration as lead petrophysicist for North America and, after a 2-year assignment as ExxonMobil's Geoscience Best Practices Coordinator, assumed his current role as ExxonMobil's global Formation Evaluation Skill Area Coordinator.

Hezhu Yin holds a B.S. in Geophysical Engineering from the University of Petroleum, China, and a Ph.D. in Geophysics from Stanford University. He taught at Da-Qing Petroleum Institute for 5 years as an Assistant Professor and 3 years at Lamont Doherty Earth Observatory of Columbia University, New York. He joined Exxon Production Research Company in 1995 (now ExxonMobil Upstream Research Company). His research interests and specialties include resistivity, nuclear, and sonic-tool response modeling and inversion, log-seismic integration and interpretation, rock physics, and 3-D seismic and petrophysical properties inversion.

Bob Brackett is currently a consultant in McKinsey & Company's Petroleum Practice. Prior to joining the firm, Brackett spent 8 years at ExxonMobil in a number of assignments in research, development, and exploration. Bob holds a Ph.D. in Earth & Planetary Sciences from Washington University — St. Louis, an MBA from Rice University, a B.S. in Geophysics and a B.A. in Astronomy from the University of Texas — Austin. The work presented in this publication was neither conducted on McKinsey's behalf nor does it reflect the firm's perspective.

Yuehui Xiao holds a B.S. degree in Geochemistry from the University of Science & Technology of China and a Ph.D. in Mineralogy/Geology from University of Illinois. He joined Exxon Production Research Company in 1996 and has worked on a variety of research, exploration, and development projects. Xiao is currently on assignment to Qatar RasGas as a petrophysicist.

Angel G. Guzmán-Garcia has a Ph.D. degree in Chemical Engineering from Tulane University. In 1990 he joined Exxon Production Research Company where he conducted laboratory and numerical research on the response of resistivity and SP tools in shaly sands, and investigated NMR petrophysical applications. In 2000, he transferred to ExxonMobil Exploration Company, where he performed petrophysical interpretation of wells in offshore West Africa and collaborated on NMR logging jobs around the globe. Currently, he works as Staff Engineer with geographical responsibility for ExxonMobil well-testing activities in Nigeria Exploration, Sakhalin, and Equatorial Guinea. Additional experience includes fluid sampling, PVT analysis, design and interpretation of PLT data, formation-tester data interpretation, and open-hole petrophysical interpretation and integration with pressure-transient analysis. Guzmán-Garcia is an editor for SPWLA and SPE Journals.

Acknowledgments

AAPG wishes to thank the following
for their generous contributions to

Petrophysical Evaluation of Hydrocarbon Pore-Thickness in Thinly Bedded Clastic Reservoirs

ExxonMobil Upstream Research Co.

Contributions are applied toward the production cost of publication,
thus directly reducing the book's purchase price
and making the volume available to a larger readership.

Table of Contents

Preface

The difficulties of obtaining accurate formation evaluation results in thinly bedded clastic reservoirs have been recognized and addressed by many people in our industry across a span of at least 50 years. The methods recommended in this publication build on the work of all these people, some of whom are acknowledged in the text by way of bibliographic reference. There are many significant contributions that we have not included in our approach. Some of these are referenced in the text and many others are not. We make no claims to completeness or uniqueness. Our integrated approach has simply evolved as our particular set of practices, and it has proven highly effective.

The work presented here originated in a research project at Exxon Production Research Company (now ExxonMobil Upstream Research Company) that was conducted during the years 1995–1998 and was first published as a series of proprietary internal reports. For this publication we have expanded on these reports and updated them to the best of our ability. The astute reader will probably find some references, especially to service company technology, that are slightly dated. We trust that these will not detract from the coherent presentation of our technical approach.

Numerous people within ExxonMobil have contributed to this work either through their own research, by providing data and opportunities for testing, or through their support and criticism. Dave Kennedy was one of the early pioneers of resistivity modeling for formation evaluation within operating companies, and was instrumental in the development of the commercially available RtBan software package that was used for most of the resistivity modeling work described here. We thank Scott Dodge, Fred Haynes, Fritz Merz, Donal Mageean, Grant Heavysege, Craig Tingey, and Greg Womack for opportunities to test many of the procedures described here. We appreciate Kevin Bohacs, Howard Feldman, Dave Jennette, Anthony Sprague, Morgan Sullivan, John Van Wagoner, and Rob Wellner for providing core descriptions and outcrop photographs illustrating thin-bed depositional environments. John Campbell provided the protocol and the photographs of procedures for acquiring and sampling cores in unconsolidated sands. Thelma LaDay and Earline Jones provided figure scanning and early document preparation. We thank the ExxonMobil operating affiliates who gave their approval to include field examples and data from numerous sites around the world.

The manuscript was improved by reviews from Mark Bowers, Michele Thomas, and Chris Rendeiro. Marianne Weaver, Karen Love, Larry Baker, and Ray Charles provided continual management support for this research and its publication.

We appreciate the assistance of Schlumberger, Baker Atlas, and Halliburton, and their representatives who provided illustrations and examples.

Finally, we thank Jack B. Thomas, former AAPG Geoscience Director, for his enthusiastic support at the inception of this volume.

Editor's Preface

In October 2003, I was asked by Quinn Passey at ExxonMobil Upstream Research Company to undertake the task of making a book out of six proprietary research reports and one published paper, the fruits of a research project conducted by the authors (myself included) during the years 1995–1998. Here, finally, is that book.

To assemble this volume I chopped, compressed, expanded, rearranged, and sometimes rewrote the words of the 7 authors. The choices of what to cut and what to keep from each report were largely mine. The authors reviewed drafts of the chapters and gave me their editorial comments. Some of these comments were incorporated and others fell by the wayside. Therefore any misrepresentations of an author's work, or any errors that may have crept into the volume, are purely the responsibility of this editor.

Some new material was added. This includes references to published work by others that is closely or tangentially related to the subjects addressed in the book. Also included is some rudimentary material on the multi-component induction logging tools that came on the scene after our original research was completed. The "thin-bed simulator" spreadsheet that is referenced in Chapter 1 and located at http://search.datapages.com/data/open/archie01.xls is my doing. Again, if there are any complaints about any of this material, the buck stops here.

I thank my co-authors for their assistance in updating the text and reviewing the drafts, and I thank Beverly Molyneux of AAPG for her thorough and enthusiastic work on the preparation of this volume.

Ken Dahlberg
PetroGeeks LLC
September, 2005

**Deep-water (Basinal) Turbidites,
Permian Collingham Formation:
Baviaans Syncline,
Laingsburg Karoo, South Africa**

Photo by Anthony Sprague

1 The Clastic Thin-bed Problem

Overview

Introduction

This chapter describes the difficulties associated with formation evaluation in thinly bedded sandstone reservoirs and outlines the existing technology that has been applied to these difficulties. Chapter 2 introduces our integrated approach to evaluating hydrocarbon pore-thickness in thinly bedded clastic reservoirs.

Many terms are introduced in these two chapters through examples and informal definitions: for example, *earth models*, *bed types*, *logging tool forward models*, *convolution*, *inversion*, *statistical earth models*, and *Monte Carlo inversion*. These topics are discussed more thoroughly in Chapters 6 through 10, and the integrated approach is described in detail in Chapters 11 through 13.

Contents

DOI: 10.1306/1157784A13220

Well-log Resolution and Hydrocarbon Pore-thickness

Well-log vertical resolution

Most oil and gas reservoirs contain some recoverable hydrocarbons in beds with thicknesses below the resolution of conventional logging tools. Conventional log analysis methods, including shaly sand methods, underestimate hydrocarbon pore-thickness in such beds.

The practical impact of this vertical-resolution effect ranges from negligible to severe, depending on the bed-thickness distribution of the reservoir. Figure 1.1 illustrates two reservoirs where the impact is severe.

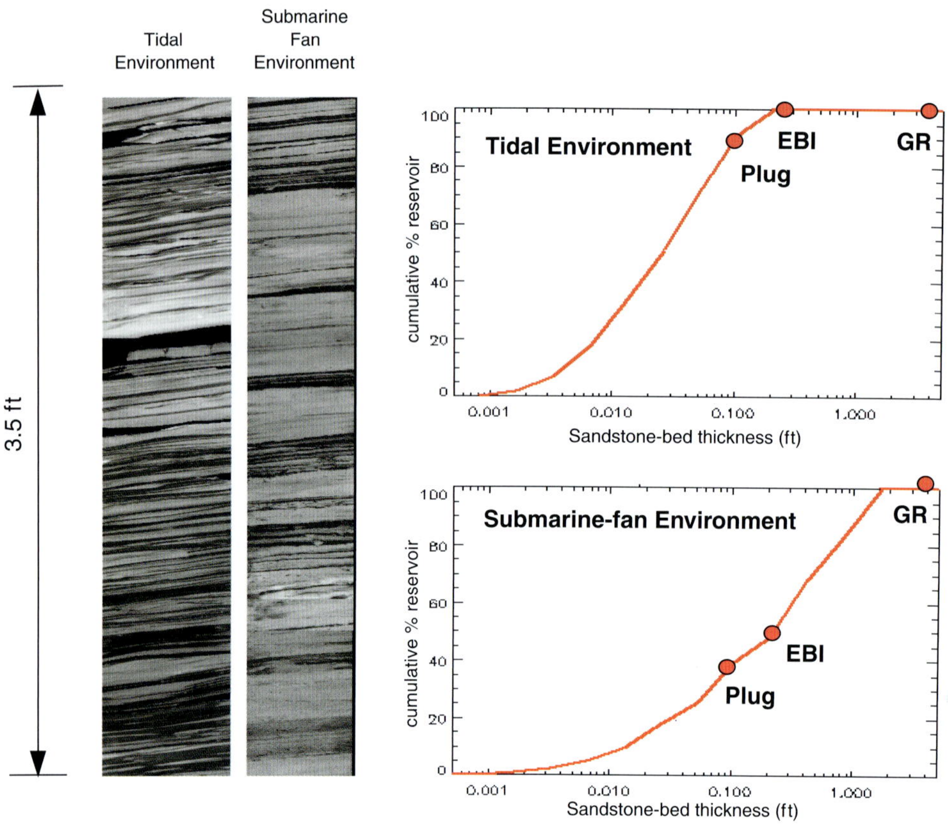

Figure 1.1. Ultra-violet-light core images from two thinly bedded reservoir intervals and their associated bed-thickness-distribution plots showing the approximate resolutions of core plugs and gamma-ray (GR) and electrical borehole image (EBI*) logs.

**Note: We use a non-standard name and acronym (electrical borehole image, EBI) to refer generically to any one of a collection of micro-electrical borehole imaging tools. For reference to specific imaging tools, see Chapter 9.*

Figure 1.1 exemplifies the magnitude of the thin-bed problem in reservoir intervals from two different geological settings. The first example is from a tidal flat environment. From the plot of cumulative percent reservoir vs. (reservoir) sandstone-bed thickness it is evident that all bed thicknesses are below the vertical resolution of the gamma ray (GR) log and even the high-resolution electrical borehole image log (EBI). For this example, approximately 80% of the reservoir volume in this interval occurs in beds below standard core-plug diameter (1 in. [2.5 cm]).

The second example in Figure 1.1 is from a distributary lobe complex in a deep-water Gulf of Mexico reservoir. All of the reservoir volume occurs in beds thinner than GR log resolution and approximately 50% of the volume occurs in sandstones thinner than EBI resolution. For this example, approximately 30% of the reservoir volume occurs in beds below standard core-plug diameter (1 in. [2.5 cm]).

In these pages, the vertical resolution of a logging tool is defined as the thickness of the thinnest bed in which a true reading can be obtained.

Common logs may be ranked from lowest to highest vertical resolution as follows. The absolute resolution of a given log varies depending on the specific tool's intrinsic resolution, the data sampling rate, logging speed, and data processing methods.

- Spontaneous potential
- Deep resistivity
- Gamma ray
- Bulk density, neutron porosity, acoustic
- Very shallow resistivity, e.g., microspherically focused log (MSFL)
- Dipmeters and electrical borehole image (EBI) logs

Since the deep resistivity is a key log in evaluating reservoir hydrocarbons, its vertical resolution (generally 2 ft [.6 m] or more) defines the lower limit of bed thicknesses below which the *thin-bed problem* begins to be significant.

Definition: HPT and OHIP

An important aspect of most reservoir assessments is the determination of original hydrocarbons-in-place (OHIP). For an oil reservoir, the original oil-in-place (OOIP) at surface conditions can be determined from the *volumetric equation:*

$$\begin{aligned} \text{OOIP} &= A \cdot h \cdot \phi \cdot (1-S_{wi}) \cdot 7758 \cdot (1/B_{oi}) \\ &= A \cdot \text{HPT} \cdot 7758 \cdot (1/B_{oi}) \end{aligned} \tag{1.1}$$

where OOIP = original oil-in-place [stock tank barrels, stb];

A = gross reservoir area [ac];
h = average oil-bearing rock thickness [ft];
ϕ = average total porosity of oil-bearing rock [frac];
S_{wi} = average total initial water saturation of oil-bearing rock [frac];
B_{oi} = the average initial oil formation volume factor [reservoir barrels per stock tank barrel, rb/stb]; and

$$\text{HPT} = h \cdot \phi \cdot (1-S_{wi}) = \textit{hydrocarbon pore-thickness} \text{ [ft]} \tag{1.2}$$

The quantity we call HPT is often referred to in the petrophysical literature as HPV (hydrocarbon pore-volume). Strictly speaking, however, this quantity is a measure of thickness rather than volume.

Conventional log derivation of HPT

In solving Equation 1.1, the petrophysicist is involved in the determination of h, ϕ, and S_{wi} from an integrated analysis of routine core analysis data, special core analysis data, and well logs. The general process for conventional, depth-by-depth log analysis is described below.

The *incremental hydrocarbon pore-thickness* (HPT_{inc}) for a single depth increment in a well is defined by the product of increment thickness (h_{inc}), log-derived porosity (ϕ_{inc}), and log-derived hydrocarbon saturation ($1-S_{winc}$):

$$HPT_{inc} = h_{inc} \cdot \phi_{inc} \cdot (1-S_{winc}) \tag{1.3}$$

The log-derived porosity typically is core-calibrated, and represents the total interconnected pore volume at reservoir conditions (e.g., Boyle's Law porosity at net overburden pressure).

The water saturation may be either resistivity-based or capillary pressure-based (e.g., from a function relating porosity and/or permeability and the height above free-water level to water saturation). In either case, the value is representative of a total water saturation that includes clay-bound, capillary-bound, and free water.

Total hydrocarbon pore-thickness (HPT) is then calculated by summing up the incremental HPT over the interval of interest:

$$HPT = \Sigma\,(HPT_{inc}) \tag{1.4}$$

In conventional log analysis, this summation is performed commonly on only those intervals that have been interpreted as *net reservoir* or *net pay* by the petrophysicist or geologist. Differentiating net from non-net is accomplished typically by applying one or more *cutoffs* to continuous, log-derived quantities such as shale volume, porosity, and water saturation. In a hypothetical example, a petrophysicist may exclude increments with shale volume exceeding 50%, porosity less than 12%, and water saturation greater than 80%. The implications of this practice in thinly bedded reservoirs are significant, and are discussed in the following pages.

Note: Interval averages and HPT

The average thickness, porosity, and water saturation used in the expression for HPT (Equation 1.2) are defined as follows:

$$h = \Sigma\, h_{inc} \tag{1.5}$$

$$\phi = (\Sigma\, h_{inc} \cdot \phi_{inc})/(\Sigma\, h_{inc}) \tag{1.6}$$

$$S_{wi} = (\Sigma\, h_{inc} \cdot \phi_{inc} \cdot S_{winc})/(\Sigma\, h_{inc} \cdot \phi_{inc}) \tag{1.7}$$

All summations in these expressions are taken over the same set of reservoir increments. It is important to recognize that average porosity must be thickness-weighted (as in Equation 1.6) and average saturation must be porosity and thickness-weighted (as in Equation 1.7). If simple, unweighted averages are used in Equation 1.2, it will produce a different (and incorrect) answer for HPT than Equation 1.4.

Vertical resolution and HPT

Figure 1.2 and Table 1.1 illustrate the effect of logging-tool vertical resolution on HPT calculations for a 2-ft-thick [.6-m-thick] bed. Figure 1.2 shows synthetic gamma ray (GR), array induction of 1-ft [.3-m], 2-ft [.6-m], and 4-ft [1.2-m]-deep resistivity (AO90, AT90, and AF90), and bulk density (RHOB) logs for a 2-ft-thick [.6-m-thick] layer of sandstone with thick shales above and below. In each track, the squared curves are the true parameter values for the bed. Note that the bed is resolved by the RHOB and AO90 logs, but not by the AT90, AF90, or GR logs. The impact on HPT for various log combinations using the standard Archie equation is illustrated in Table 1.1.

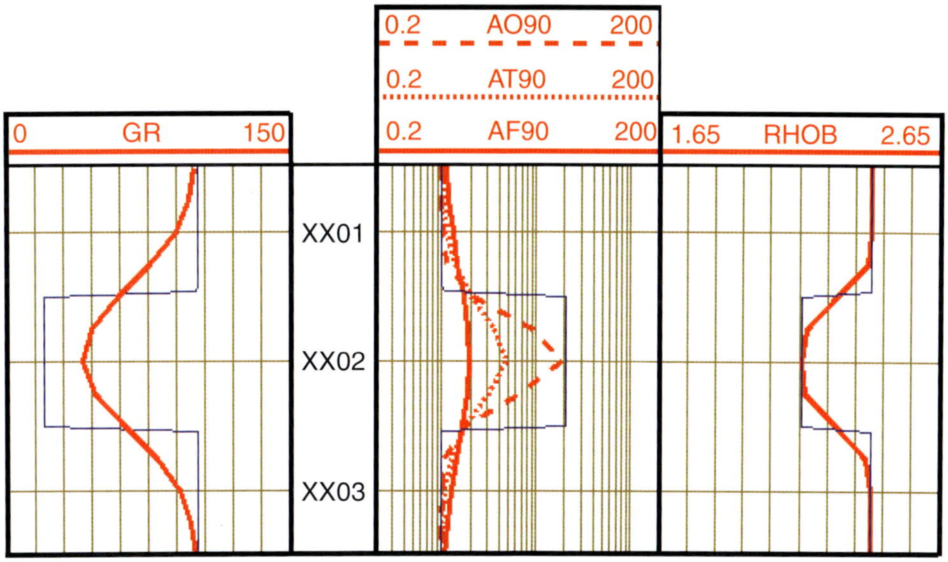

Figure 1.2. Synthetic logs for a 2-ft-thick [.6-m-thick] layer of sandstone.

Table 1.1. HPT calculations in a 2-ft-thick [.6-m-thick] layer of sandstone. DPHI = density-log porosity; R_t = true formation resistivity; POR = formation porosity.

Archie analysis inputs	Apparent HPT (ft)	Percent of true HPT
AF90, DPHI	0.32	60%
AT90, DPHI	0.37	70%
AO90, DPHI	0.42	79%
R_t(model), DPHI	0.48	91%
R_t(model), POR(model)	0.53	100%

Cutoffs and net sand in thin beds

As previously described, in conventional log analysis the discrimination of net reservoir from non-net is accomplished typically by applying one or more *cutoffs* to continuous, log-based results (e.g., shale volume, clay volume porosity, and/or water saturation). As an example, if the clean sands in a reservoir have an average gamma-ray (GR) reading of 20 API units and the shales average 100 API, then the midpoint value (60 API) might be used as a cutoff, so that each depth where GR is less than 60 is designated as sand.

In vertical wells, for beds thicker than about 20 ft [6 m] errors resulting from the application of cutoffs to determine net pay from log data may be insignificant or negligible. This is because the thickness that is affected by vertical-resolution limitations at bed boundaries is small relative to the total thickness of the reservoir.

In thinly bedded reservoirs, log resolution limitations alone can contribute to significant errors in HPT even if net reservoir is correctly determined. This was illustrated by the example of Figure 1.2, where log-derived HPT calculations were significantly low even when the shale cutoff identified the top and base of the sand accurately.

The use of cutoffs in thinly bedded reservoirs can introduce additional large errors in HPT through the incorrect determination of net sand. The examples below illustrate the magnitude of these errors.

Note:
Thin-bed simulator

Synthetic log model calculations can be useful to study the effects of thin bedding on conventional log-based HPT calculations. To facilitate these calculations, a thin-bed simulator *has been developed in the form of an Excel® spreadsheet, and is located at http://search.datapages.com/data/open/archie01.xls. The* thin-bed simulator *generates random thinly bedded sand-shale reservoir models and uses simple convolution filters to estimate log responses across these model intervals. These synthetic logs then form the input for several methods of conventional log analysis (Archie with cutoffs, Dual-water with cutoffs, Archie with no cutoffs, and Dual-water with no cutoffs). The HPT resulting from these analyses may be compared to the actual HPT for each synthetic model. The examples below, illustrating the effects of the use of cutoffs and the overall performance of conventional log analysis models, were developed using this* thin-bed simulator.

Figure 1.3 shows simulated log data across a 20-ft [6-m] thinly bedded reservoir interval. The simulated sands and shales are shown in yellow and green, respectively, in the track between the gamma ray and resistivity curves. Track 1 shows the gamma ray in blue. The cutoff used to determine net sand is the red dashed line halfway between the clean-sand and shale endpoints. The resulting calculated net sand intervals are shaded tan. The resistivity log, simulated by applying a convolution filter to the thin-bed conductivity values, is shown in Track 2; the simulated density and neutron logs are in Track 3.

For this example, the actual net sand fraction is 50% and the value determined from the gamma-ray cutoff is 39%. It is evident that the location of the net-sand intervals identified by the cutoff on the gamma-ray log bears little resemblance to the actual distribution of net sand over the reservoir interval.

By using the *thin-bed simulator,* it is a straightforward process to generate a large number of simulated reservoir intervals with varying sand fractions, and to compare the actual and the cutoff-derived net sand fraction. Figure 1.4 shows such a comparison for a set of 130 simulations with sand fraction ranging from 10% to 100%. (See the *note* on page 10 for the details and assumptions underlying these simulations.)

Figure 1.4 shows that net sand tends to be overestimated by the cutoff when above 50% and underestimated when less than 50%. These synthetic results should not be over-generalized. Nonetheless, the pattern illustrated in Figure 1.4 illustrates clearly the kinds of errors that can be expected when log cutoffs are used to identify net sand in reservoirs where beds are thinner than log resolution.

The next section illustrates how these errors in net-to-gross (N/G) contribute to errors in HPT with conventional log-analysis methods.

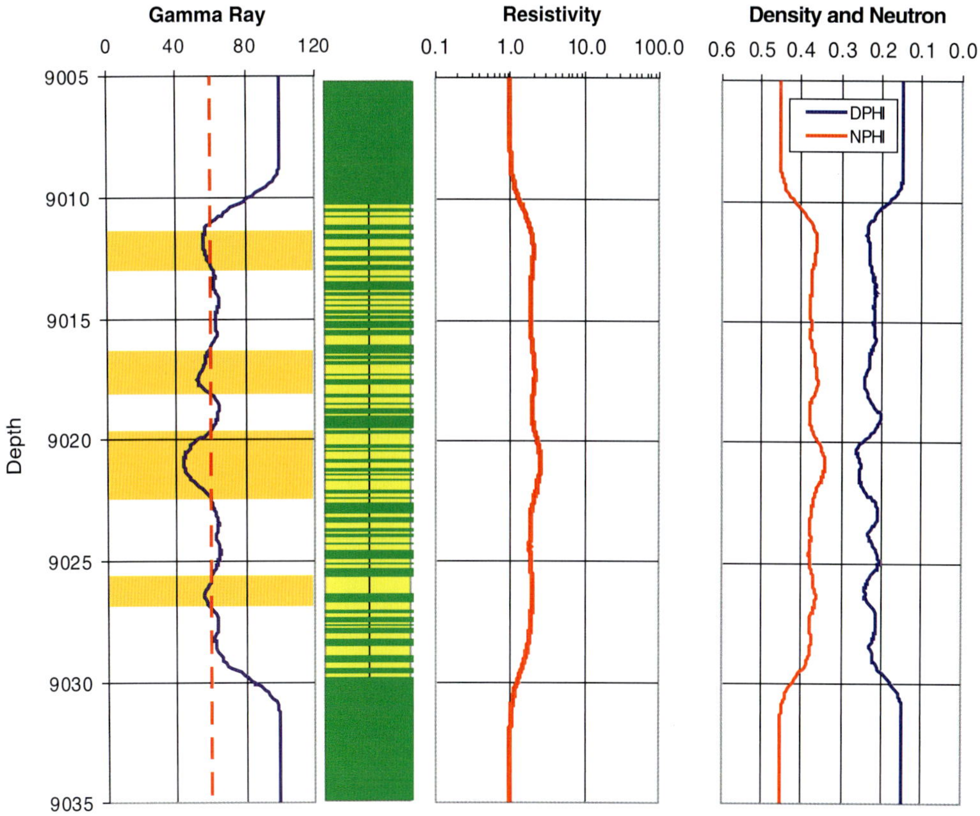

Figure 1.3. Synthetic thin-bed earth model and corresponding log response.

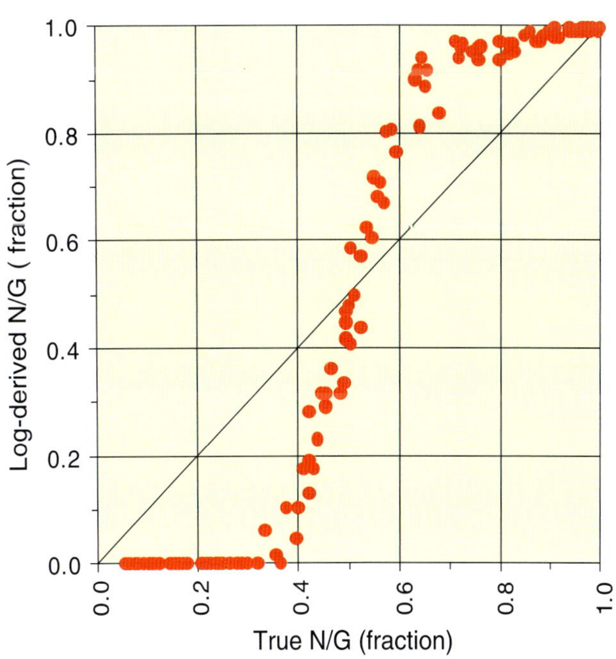

Figure 1.4. Comparison of actual and GR-derived net sand fraction (N/G) for 130 thin-bed reservoir simulations.

Archie and Shaly Sand Methods

Introduction

Regardless of analysis method, the net error in hydrocarbon pore-thickness for a thinly bedded reservoir is a combination of errors in all three factors in the HPT equation: the net sand thickness, the sand porosity, and the sand water saturation.

- As illustrated in Figure 1.4, the net sand thickness estimated by conventional cutoff methods is likely to be too high if actual net sand is greater than 50%. It is likely to be too low if net sand is less than 50%.
- Sandstone porosity is affected by porosity-tool resolution, so the porosity measured in the assumed net sand interval is an average of sandstone and shale log readings. With the density log, assuming shale porosity is lower than sandstone porosity, the likely result is that the calculated sandstone porosity is too low.
- Assuming shale resistivity is lower than sand resistivity, the measured resistivity over the assumed net sand interval is significantly reduced by the intervening thin shales. Water saturation derived from an Archie calculation, therefore, is likely to be too high.

Shaly sand analysis

Even though thinly bedded sandstone formations have historically been called *shaly sands* in the petrophysical literature, the standard shaly sand log analysis models (e.g., Waxman-Smits and Dual-water; see Worthington [1985] for an overview) do not correctly account for the effect of thin bedding on log responses. These techniques were developed to address the effects of dispersed clay in sandstones, rather than macroscopically interbedded sandstones and shales. The electrical effects of these two modes of clay (or shale) distribution are significantly different. Nonetheless, one might expect that the application of these dispersed-clay shaly sand models to a thinly bedded reservoir would move the calculated water saturation in the correct direction relative to the Archie model. This is indeed the case, as the following examples illustrate. These examples also show that this "correction" is not calibrated to the specific effects of thin bedding.

Archie and Dual-water examples

The most concrete way to understand exactly how Archie and shaly sand techniques perform in thinly bedded formations is to make a series of synthetic model calculations, where the true formation water saturation is known and can be compared with saturation values derived by different log analysis techniques. The *thin-bed simulator* described on page 6 provides a facility to make such calculations using the Archie and Dual-water models, and to test the effect of gamma-ray net-sand cutoffs on each method.

Table 1.2 shows the results of Archie and Dual-water calculations applied to the synthetic dataset illustrated in Figure 1.3. See the *note* on page 10 for details of these calculations and the underlying assumptions.

Table 1.2. Archie and Dual-water (D-W) analyses.

Reservoir Property	Case				
	Model values	Archie with cutoff	D-W with cutoff	Archie, no cutoff	D-W, no cutoff
Net-to-Gross (N/G, %)	50%	39%	39%	100%	100%
Sand porosity (frac)	0.30	0.24	0.24	0.22	0.22
Sand S_w (frac)	0.10	0.57	0.47	0.64	0.51
HPT (ft)	2.71	0.79	0.98	1.60	2.18
Percent of actual HTP	100%	29%	36%	59%	80%

Here are some observations drawn from Table 1.2:

- When Archie water-saturation analysis is combined with cutoff-derived net sand, the calculated HPT is less than one-third of the true value. This large error is the combination of errors in net sand thickness, sand porosity, and sand water saturation.
- The Dual-water analysis with cutoff produces a slight improvement in water saturation but HPT is still just over one-third of its true value.
- When HPT is accumulated across the whole reservoir interval (i.e., when cutoffs are not used to restrict the accumulation to "net sand"), results are improved but are still significantly low for both the Archie and Dual-water analyses.
- Each of these approaches produces a log-derived water saturation that is much higher than the actual sand water saturation.

With the *thin-bed simulator* we can make a series of synthetic-model calculations like that illustrated in Figure 1.3, and collect results like those shown in Table 1.2 across a wide range of reservoir net-sand fractions. The graph in Figure 1.5 shows the comparison of calculated to true HPT values for 130 such model calculations. The horizontal axis shows the true (model) net sand fraction for each model, and the vertical axis shows calculated HPT as percent of model HPT for 4 methods: Archie and Dual-water, with and without cutoffs. Model net sand fraction ranges from 10% to 100%. (See the *note* on page 10 for details and assumptions underlying these simulations.)

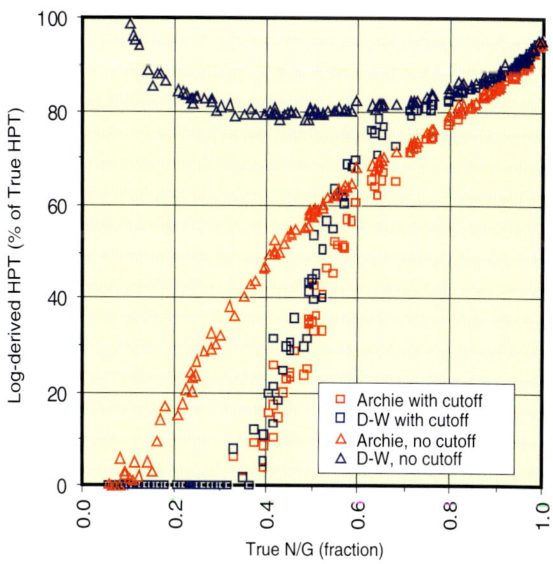

Figure 1.5. Log-derived HPT for 130 simulated thin-bed reservoirs.

Here are some observations drawn from Figure 1.5.

- The conventional Archie or Dual-water analysis does not begin to approach the true HPT except when N/G exceeds 90%. In most cases the errors in HPT are 20% or greater.
- The use of cutoffs on standard-resolution logs to identify "net sand" in thinly bedded reservoirs significantly degrades the accumulated HPT results, for both the Archie and Dual-water methods.
- The Dual-water model produces a "correction," which moves the water saturation, and thus HPT, in the right direction. In this example, the "correction" is magnified at higher shale fractions (lower N/G). Across most of the range of potentially productive reservoir N/G, the HPT accumulated without cutoffs is around 20% low.

Note: Modeling assumptions in thin-bed simulator

The synthetic formations generated by the thin-bed simulator *comprise a binary system of sandstone and shale beds having a minimum thickness of 0.1 ft [.03 m]. Table 1.3 summarizes the properties of the sand and shale beds used to generate the examples shown in Figures 1.3–1.5.*

Table 1.3. Sand and shale parameters for thin-bed simulator examples.

Parameter	Units	Value	
		Sand	Shale
density porosity	fraction	0.30	0.15
neutron porosity	fraction	0.29	0.45
water saturation	fraction	0.10	1.00
Rw (brine resistivity)	ohm m	0.040	0.023
m (cementation exponent)	none	2.00	2.00
n (Archie exponent)	none	2.00	
Gamma Ray	GAPI	20	100
GR cutoff for net sand	GAPI	60	
resistivity	ohm m	44.44	1.00
conductivity	mmho/m	22.5	1000.0

In a simulation, each 0.1-ft [.03 m] interval is assigned randomly as sandstone or shale and receives the corresponding properties listed in Table 1.3. Gamma-ray, conductivity, density porosity, and neutron porosity logs are then generated using simple convolution filters (Chapter 10).

Cutoff-derived "net sand" occurs where the gamma-ray log is below the cutoff value listed in the table.

For both Archie and Dual-water analyses, reservoir porosity is assumed equal to density porosity.

In the Dual-water analysis, the bound-water resistivity (R_{wb}) is determined by apparent water resistivity (R_{wa}) analysis in the thick shale. For the current examples, this value is shown in Table 1.3 as 0.023 ohm m. The bound water saturation (S_{wb}) is taken to be 0.0 in the sandstones and 1.0 in the shales. At each depth, S_{wb} can then be calculated from the GR-derived shale volume and the sandstone and shale porosities, which are assumed known.

A Brief Overview of Thin-bed Technology

Introduction

The problems identified above have been well known in formation evaluation for many years, and the proposed solutions have grown more sophisticated over time. Here we present a brief conceptual and historical introduction to some of technology related to the thin-bed problem. Most of these topics will be covered in greater detail in subsequent chapters.

Logging tool forward modeling and inversion

A detailed understanding of how each logging instrument responds to the bedding geometry of a thinly bedded formation is obviously of fundamental importance in determining how to extract accurate HPT estimates from logs in such formations. A logging tool *forward model* is a computational algorithm that starts with a description of bedding geometry and bed properties (that is, an *earth model*, such as the sandstone and shale beds illustrated in Figure 1.3 and described in Table 1.3) and produces a computed approximation to what the logging tool would actually measure in such a formation. Since the exponential growth in available computing power that began in the 1980s, the use of such forward models has become an indispensable tool in understanding resistivity logs [Anderson et al., 1989; Kennedy, 1995].

A forward model for a logging tool may be a highly complex and detailed algorithm that solves the fundamental equations governing the tool's physics in either a complete or a simplified geometric configuration [Anderson and Gianzero, 1983; Anderson, 1986]. Such a model may produce a highly accurate approximation to the tool's response under a wide range of conditions. On the other hand, there is a class of much simpler forward models that make a *linear approximation* to the tool's response in simple orthogonal bedding geometries like those in Figure 1.3. These models, called *convolution models* or *convolution filters*, have been widely used in modeling induction logs [Moran, 1982] as well as other common logs [Looyestijn, 1982].

When a forward model is available for a given log measurement, it becomes possible to consider the *inversion* problem: given the measured log, what is the set of bedding geometries and bed properties that could have produced that log? One solves a form of this problem every time one makes an estimate of HPT using logs. We will see that, when we are dealing with beds thinner than log resolution, the inversion problem is *non-unique*: in other words, there is no single solution for a set of thin beds and their properties that would produce the given log(s) [Yin, 2000].

A fundamental conductivity relationship

Most of the work on the thin-bed problem has focused on resistivity logs, since the resistivity-derived water saturation is very strongly affected by the presence of interbedded shales. A large fraction of this work has made use of the following simple relationship:

$$\sigma_{\underline{av}} = v_{sh} \cdot \sigma_{sh} + (1 - v_{sh}) \cdot \sigma_{sd} \qquad (1.8)$$

In Equation 1.8, $\sigma_{\underline{av}}$ is the average formation conductivity, measured parallel to the bedding planes, across some finite interval perpendicular to the bedding planes; σ_{sh} is the average conductivity of the interbedded shales; σ_{sd} is the average conductivity of the interbedded sandstones; and v_{sh} is the fraction of shale beds within the averaging interval.

It has been demonstrated by published forward-model calculations [Anderson, 1986] that Equation 1.8 is a good representation of the conductivity response of an induction logging tool when the borehole is perpendicular to the bedding planes and invasion is not significant. A slightly more complex equation that accounts for dip effects (to be discussed later) has been similarly validated by model calculations [Anderson et al., 1988].

Equation 1.8 explains the large impact of interbedded shales on measured resistivity in thinly bedded reservoirs. For example, the average resistivity-log value in the reservoir interval of Figure 1.3 is only about 2 ohm m, despite the formation being 50% sandstone and each sandstone bed having 44 ohm m resistivity (Table 1.3).

Methods using standard logs

The earliest proposed solutions to the thin-bed problem [e.g., Poupon et al., 1954] relied on solving Equation 1.8 for the sand conductivity as in Equation 1.9.

$$\sigma_{sd} = (\sigma_{av} - V_{sh}\sigma_{sh})/(1 - V_{sh}) \qquad (1.9)$$

More recent variations [e.g., Van den Berg et al., 1996] have generalized Poupon's original laminated-sand analysis to account for the effect of relative dip. In all of these approaches, the shale fraction (V_{sh}) is estimated using some combination of shale-sensitive logs such as the gamma ray and the density-neutron pair. The shale conductivity (σ_{sh}) is taken from a resistivity log reading in a thick shale. The "corrected" sandstone conductivity, σ_{sd}, is then used to calculate sandstone–water saturation.

Examining Equation 1.9, one can see that errors in V_{sh} may yield potentially significant errors in σ_{sd} and thus in sand–water saturation. This sensitivity can be tested using the example of Figure 1.3. Taking the sandstone and shale conductivities from Table 1.3 and assuming an actual sandstone fraction of 0.50, we can plot the values of sand resistivity (R_{sd}) derived from Equation 1.9 for various values of V_{sh} near the correct value of 0.50.

The bold red diamond in Figure 1.6 represents the correct solution for R_{sd} when V_{sh} is 0.50. Moving away from the red symbol, each dark blue symbol represents an additional 1% error in V_{sh}. The errors in R_{sd} are posted on the plot for the first 1% errors above and below the correct value of V_{sh}.

Figure 1.6 demonstrates that, in some cases, the sensitivity of this classic method to errors in V_{sh} is extreme and unacceptable. But we will see below, in the exposition on Volumetric Laminated Sand Analysis [Chapter 12], that Equation 1.8 can still be useful if we do not attempt to solve it for σ_{sd}.

Figure 1.6. Apparent sandstone resistivity (R_{sd}) versus V_{sh}.

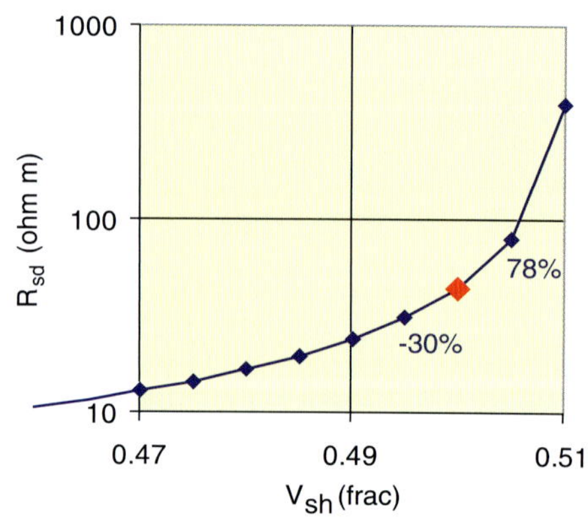

High-resolution earth models

Several specialized high-resolution logs, such as image logs, dipmeters, and Schlumberger's Electromagnetic Propagation Tool, provide the possibility of resolving bedding down to a scale of about 2 in. [5 cm]. These logs may then be used to construct *high-resolution earth models.* Again, the yellow and green sandstone–shale geometry in Figure 1.3 is a synthetic example of such a model.

High-resolution earth models are used in several ways to improve the derivation of HPT in thinly bedded reservoirs. One published approach uses convolution filters to combine the high-resolution earth model with standard-resolution log data and solve for an approximate high-resolution resistivity profile [e.g., Allen, 1984; Ruhovets, 1990]. This approach requires several simplifying assumptions and while it may be useful in particular circumstances, we do not pursue it further here.

A more general approach is to use the high-resolution earth model as the basis for forward-modeling the log responses. Historically, most of the attention here has been on the modeling of resistivity tools [e.g., Anderson, 1986; Anderson et al., 1989; Kennedy, 1995; Bergslien, et al., 2000]. There has been at least one commercial approach that used convolution filters in conjunction with a high-resolution earth model to reconstruct both resistivity and porosity logs at the resolution of the earth model. This was the Schlumberger SHARP processing method [Boyd et al., 1995; Serra and Andreani, 1991].

Our high-resolution approach [Chapter 11] uses convolution filters in simple cases. In more complex cases we use convolution filters for porosity logs and more detailed forward models for resistivity logs. As the example of Figure 1.2 shows, the most accurate high-resolution evaluation of HPT requires correcting the resolution of both the porosity and the resistivity data. The method of using a high-resolution earth model as the foundation for forward modeling reduces some of the uncertainty associated with the non-uniqueness of inversion when the beds are thinner than the intrinsic resolution of the logs. However, it is important to remember that there are practical limits to this approach. Frequently one must deal with formations where at least some of the beds cannot be resolved by any log measurement, as Figure 1.1 illustrates.

Nuclear Magnetic Resonance logging

The Nuclear Magnetic Resonance (NMR) log provides a volume-averaged measurement that indicates the presence or absence of moveable (free) fluids in the logged interval. NMR logs are often the first indication of the possibility of producible hydrocarbons in an interval that looks extremely shaly on conventional logs [Akkurt et al., 1997]. The NMR produces low-resolution log curves that cannot be used to quantify individual thin beds, but under ideal conditions NMR can provide a direct measurement of total HPT over a thinly bedded interval.

Electrical anisotropy

Rocks that have a preferred orientation, or layering, on a scale finer than the scale of resistivity log measurement will likely exhibit *electrical anisotropy*, which is a dependence of the measured conductivity (or resistivity) on the direction of current flow through the rock. Thus, referring to the synthetic formation of Figure 1.3, the conductivity measured by currents parallel to the sandstone and shale bedding planes satisfies Equation 1.8. This is the conductivity (and associated resistivity) measured by conventional induction and focused-current logging tools when the borehole is perpendicular to the bedding planes. As discussed above, this *parallel resistivity* measured in a thin-bedded reservoir is very much lower than the resistivity of the sand beds and is extremely sensitive to the high-conductivity shales.

On the other hand, the resistivity measured by currents transverse (or perpendicular) to the bedding planes satisfies Equation 1.10.

$$R_{perp} = V_{sh} \cdot R_{sh} + (1 - V_{sh}) \cdot R_{sd} \quad (1.10)$$

For the example of Figure 1.3, this resistivity would average about 23 ohm m, or half the sand resistivity, compared to the 2 ohm m average measured in the parallel direction and described by Equation 1.8. Thus, the *transverse resistivity* is much more sensitive to the resistive sands than is the parallel resistivity.

The effect of anisotropy on electrical log measurements has been known and understood in theoretical terms for many years [Moran and Gianzero, 1979]. More recently, the prevalence of directional drilling has led to increased interest in finding practical analytical solutions for anisotropic effects [Klein, 1993; Hagiwara, 1995], and in developing resistivity tools that can measure these effects directly.

Multi-component (triaxial) induction tools

Since the late 1990s there have been intensive efforts to develop induction logging tools capable of measuring all the directional components of the formation *conductivity tensor* [Fanini et al., 2001; Kennedy et al., 2001]. If it were possible to accurately measure the resistivity transverse to bedding in a thin-bedded reservoir, the uncertainty in the log-derived HPT would be greatly reduced. We will outline herein the principles of application of these multi-component resistivity measurements. However, there are many technical difficulties in making such measurements accurately and reliably, and at the time of writing they do not form a significant part of our integrated approach.

Summary

Most clastic oil and gas reservoirs contain some recoverable hydrocarbons in beds with thicknesses below the resolution of conventional logging tools. Conventional log analysis methods, including shaly sand methods, tend to underestimate hydrocarbon pore-thickness (HPT) in such beds.

Because of their limited vertical resolution, well logs measure an average of the properties of thin interbedded sandstones and shales. These resolution limitations can produce significant errors in HPT even if net reservoir thickness can be determined correctly. The use of cutoffs in thinly bedded reservoirs can introduce additional large errors in HPT through the incorrect determination of net sand thickness.

Standard resistivity logs measure thin hydrocarbon-saturated sandstone beds and shale beds in parallel, yielding log values much closer to the low shale resistivity than to the higher sandstone resistivity. Standard "shaly sand" log analysis models are not designed to properly correct for this effect, and the usual result is over-estimation of water saturation.

A simple parallel-conductivity equation is often used to approximate the response of standard resistivity logs in thin-bedded reservoirs. Methods have been devised to improve the estimation of sandstone water saturation by solving this equation for sandstone conductivity. These methods, which we refer to cumulatively as *conventional laminated sand analysis*, are prone to large errors because of their sensitivity to minor errors in the shale fraction.

Some of the most effective published work on the thin-bed problem has utilized a high-resolution log, such as a borehole image log, to resolve individual thin beds. Then a defined set of bed boundaries (an earth model) is used in combination with a set of forward models for the logging tools to perform an *inversion* which provides improved estimates for the true values of each log within each thin bed. There are many productive reservoirs whose beds are too thin to be defined by any high-resolution log; thus the approach of forward modeling and inversion is not universally applicable.

The Nuclear Magnetic Resonance tool and the multi-component induction tool are newer measurements that have significant potential to improve the accuracy of HPT determination in thin-bedded reservoirs.

Delta Front and Stream Mouth Bar,
Upper Cretaceous,
Blair Formation:
Baxter Basin, Wyoming, U.S.A.

Photo by Kevin Bohacs

2 A Roadmap for Evaluating Thin-bedded Clastic Reservoirs

Overview

Introduction

This chapter introduces ExxonMobil's systematic approach to evaluating hydrocarbon pore-thickness (HPT) in thinly bedded reservoirs. The material presented here is reproduced, with minor modifications, from the 2004 SPWLA paper by Passey et al.

Many terms are introduced in Chapter 1 and this chapter through examples and informal definitions: For example, *earth models, bed types, logging tool forward models, convolution, inversion, statistical earth models, Monte Carlo inversion.* These topics are discussed more thoroughly in Chapters 6 through 10, and the integrated approach is described in detail in Chapters 11 through 13.

Contents

DOI: 10.1306/1157785A13220

A Systematic Approach to Estimate HPT

Introduction: Two complementary methods

Two complementary methods have been developed to improve the estimation of hydrocarbon pore-thickness (HPT) in thinly bedded siliciclastic reservoirs. The first, generically termed the *high-resolution* method, is appropriate for reservoirs where the average thickness of sandstone and shale beds is at least 1 ft [.3 m]. The second, generically termed the *low-resolution* method, is applicable to any bed thickness and is most useful where bed thicknesses are less than 1 ft [.3 m].

In both methods, well-log responses to a postulated *earth model* are calculated and compared to measured logs. The earth model is adjusted to bring these calculated logs into agreement with the measured logs, and then HPT is calculated directly from the earth model. The two methods differ in the type of earth model (deterministic versus statistical) and in the type of log-response calculation (point-by-point versus interval average).

The High-resolution Method

Tool response modeling and inversion

A high-resolution measurement, such as an image log or high-resolution density log, can be used to define initial bed boundaries. These boundaries are the starting point to construct a detailed *earth model* comprising a set of beds, located in depth along a wellbore, together with estimates of petrophysical properties of each bed (e.g., density, gamma ray, and resistivity). This detailed earth model is then input to a set of *forward models* that predict depth-by-depth log responses.

These predicted logs are compared to the corresponding measured logs. If the agreement is poor, the earth model is adjusted and the forward models are recomputed until the predicted and measured logs come into acceptable agreement. This process of *inversion* yields an earth model that is taken to be a better representation of the true properties of the beds than the original, raw log measurements.

Synthetic example

Figures 2.1 and 2.2 illustrate the elements of the process of modeling and inversion using a synthetic formation model and synthetic resistivity data. On the left in Figure 2.1 is a (synthetic) high-resolution formation image identifying sharp bed boundaries. The resistivity log shows a synthetic "measured" resistivity, an initial inverted resistivity earth model with postulated values for each bed, and a predicted resistivity log calculated from the initial earth model. Note the mismatch between the "measured" and predicted logs, which indicates that the initial inverted earth model is not an accurate representation of the true resistivities of the formation beds.

The HPT calculated from the "measured" resistivity log is 2.62 ft [.8 m].

In the inversion step, the resistivity values of the beds in the earth model are adjusted to optimize the match between the "measured" and predicted logs. Figure 2.2 shows the final inversion result for the resistivity log based on the revised earth model. The predicted (red) log now overlays the "measured" (blue) log. The model-based inverted resistivity values (orange) are now taken to be a more accurate representation of the true formation resistivity than the "measured" log. The inverted resistivity values for the sandstone beds yield a 30% increase in calculated HPT for this interval, compared to that calculated from the "measured" log (3.43 ft versus 2.62 ft) [1 m versus .8 m].

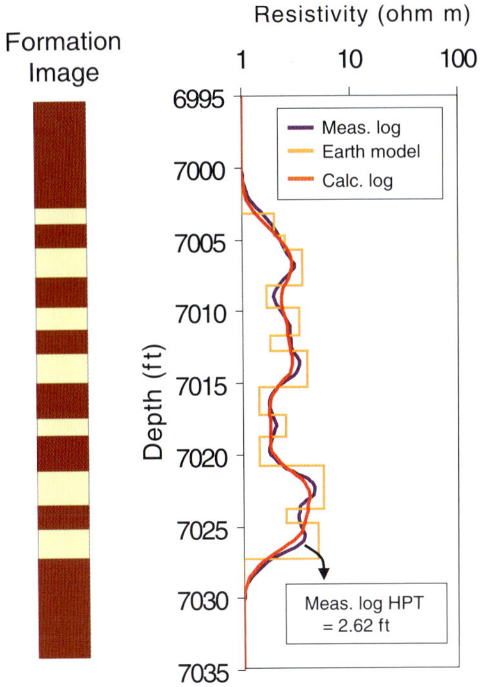

Figure 2.1. Synthetic resistivity log (blue curve) for a formation model of known properties, initial resistivity earth model (squared orange curve), and corresponding predicted log (red curve).

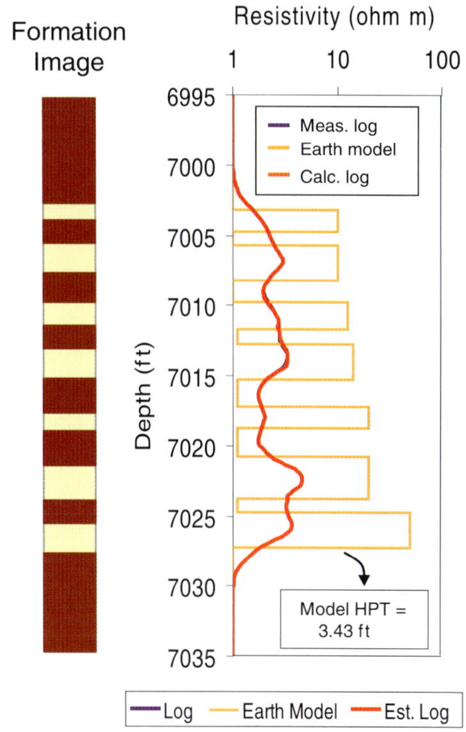

Figure 2.2. Final earth model and inverted resistivity. Note good agreement between blue (Meas.) and red (Calc.) curves.

Field example

Figure 2.3 illustrates the process of high-resolution log modeling and inversion using real data from a deep-water siliciclastic reservoir. The log on the left shows a 4-m [13-ft] section of sandstone reservoir with interbedded shale. The log on the right shows the estimates of formation properties obtained through the process of tool modeling and inversion. Modeling indicates a significant increase in sandstone resistivity and results in a 35% increase in HPT for the 4-m [13-ft] interval.

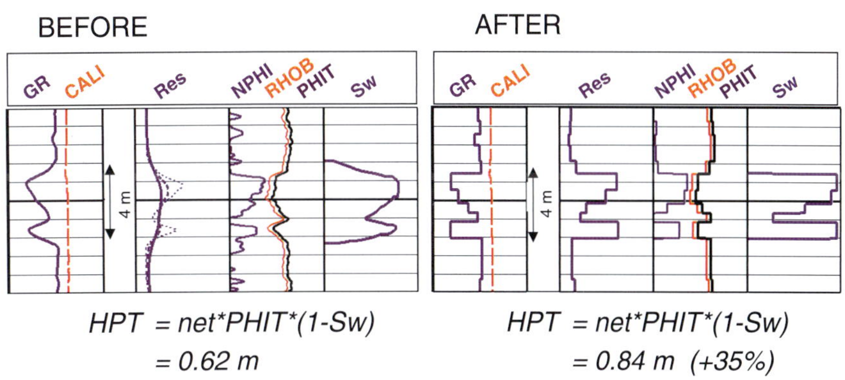

Figure 2.3. Deep-water reservoir example illustrating the impact of high-resolution log modeling on hydrocarbon pore thickness (HPT). In the BEFORE case, the resistivity curves represent shallow (dotted), medium (dashed), and deep (solid) readings. In the AFTER case, only the deep resistivity is shown.

Note, in the "AFTER" log on the right of Figure 2.3, that not only the deep resistivity but also the gamma ray (GR), bulk density (RHOB), and neutron porosity (NPHI) have been modeled and inverted to conform to a common set of bed boundaries. All three factors in the calculation of HPT (net, PHIT, and S_w) are improved through this procedure.

Conclusion

A key element in this high-resolution method is the set of *tool-response models* used to predict log responses from *earth models*. These can range from simple linear *convolution* models to highly detailed and computationally intensive two- or three-dimensional models that simulate the physics of the logging tool. Generally, we use convolution models for porosity and gamma-ray logs. For resistivity logs we apply models ranging from convolution operators to detailed one- or two-dimensional simulations, depending on the complexity of the observed log response. At ExxonMobil a series of resistivity modeling codes by service companies, software developers, and internally developed proprietary inversion codes has been integrated into a fit-for-purpose internal application. By modeling all logs required for the petrophysical evaluation using the same bed boundaries, the logs are effectively resolution-matched, and the final petrophysical assessment is not compromised by resolution "mismatch" artifacts.

The Low-resolution Method

Volumetric laminated sand analysis

Theory and experience have shown that, when bed thickness is less than a critical minimum, it is no longer possible to find a unique optimal solution using the detailed high-resolution modeling technique. To address this situation, with its resulting uncertainty, a new low-resolution technique has been devised and is called *Volumetric Laminated Sand Analysis* (VLSA).

In VLSA we begin by identifying a thinly bedded formation interval and formulating a *statistical earth model*. This statistical model comprises a set of *bed types* together with *a priori* distributions describing the range of average petrophysical properties for each bed type and the range of possible volume fractions for each bed type within the defined interval. There is no attempt to resolve individual beds; only their composite properties are described. A *realization* of this statistical model is then a set of petrophysical property values and volume fractions drawn from these *a priori* distributions. For such a realization, the interval-average well log responses are predicted using an appropriate set of forward models. A probability index is calculated, proportional to the probability that this set of predicted log averages is equal to the interval averages of the measured logs, given the uncertainty in both the log measurements and in the predictive models. This probability index is used as a weighting factor for each realization in a large sample of realizations. This sampling process is termed *Monte Carlo inversion* and its result is a new set of "inverted" probability distributions that describe the petrophysical properties of the bed types, subject to the constraints imposed by the well logs and their response models [Chapter 10].

Synthetic example

We first illustrate the VLSA technique with a synthetic example. Figure 2.4 shows synthetic log data for a 20-ft [6-m] reservoir interval comprising thin interbedded sandstones and shales in a 1:1 ratio (50% net-to-gross). The sandstone and shale bed resistivity values are assumed to be 20 ohm m and 1 ohm m, respectively. The earth model is shown in Track 2, with sands in yellow and shales in green. The dashed line in Track 1 is the gamma-ray cutoff used to identify net sand in conventional log analysis, and the "sands" identified by the cutoff are marked with tan shading. Table 2.1 (columns

1 and 2) compares the true values of sandstone net-to-gross, porosity, water saturation, and hydrocarbon pore-thickness (HPT) with the values derived by conventional log analysis. The true HPT is more than 200% higher than the conventional log-derived value.

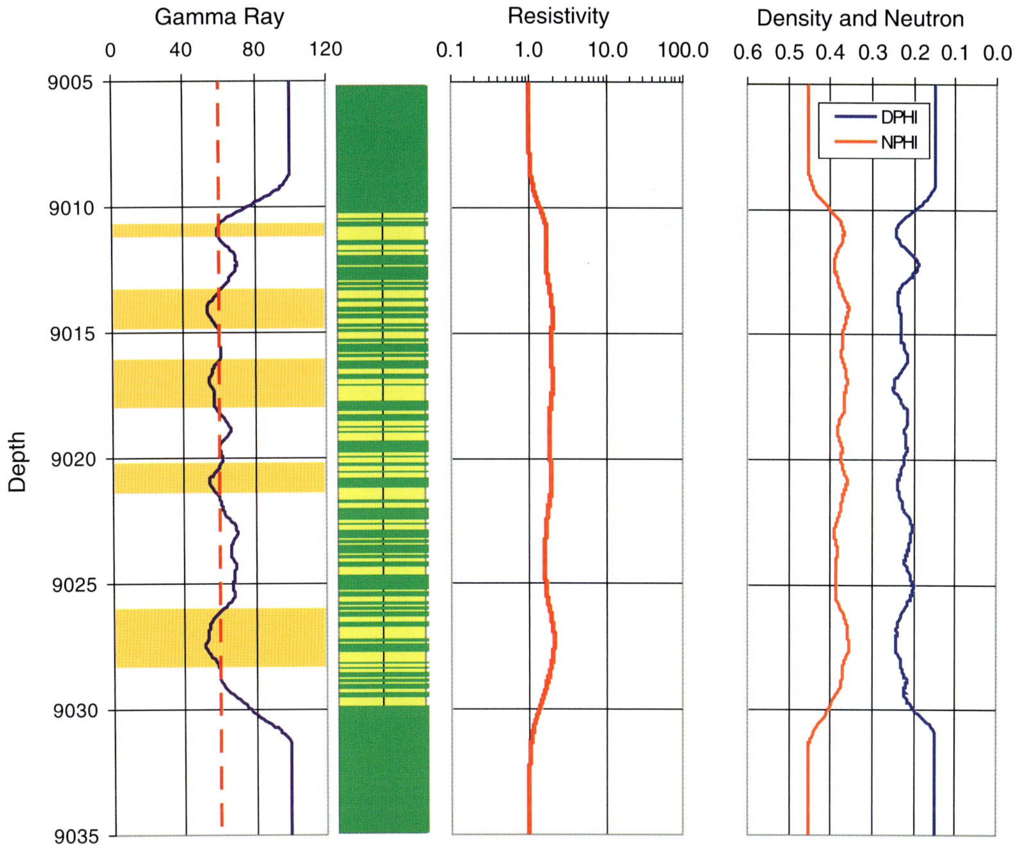

Figure 2.4. Synthetic logs for a thin-bedded formation. A comparison of the hydrocarbon pore-thickness (HPT) results from different methods for this example is shown in Table 2.1.

Table 2.1. Comparison of true reservoir parameters with those obtained by conventional log analysis and VLSA.

	True Value	Conventional Log	Statistical (VLSA)
Net-to-gross (%)	50	39	58
Porosity	30	23	29
S_w (%)	15	61	28
HPT (ft)	2.5	0.7	2.4

Figure 2.5 shows the main *a priori* and inverted property distributions for the sandstone bed-type in the synthetic example of Figure 2.4. In the VLSA analysis illustrated in Figure 2.5, the *a priori* distributions for sand fraction, porosity, and water saturation are broad uniform distributions reflecting little or no *a priori* knowledge of these properties. The Monte Carlo inversion was constrained by the interval averages of the conductivity, density, and gamma-ray logs. Note the accuracy of the inverted mean values in comparison to the values obtained by conventional log analysis (Table 2.1).

The standard deviations in Figure 2.5 reflect the degree to which each parameter is constrained by the well logs — thus, they reflect the uncertainty in the results of the analysis. In particular, the standard deviation of water saturation (S_w) is almost 50% of its mean value, indicating that S_w is only weakly constrained. It has been noted previously [Van den Berg et. al., 1996] that, in thin-bedded reservoirs with high shale fraction, the resistivity log is more sensitive to the shale fraction than to water saturation. In this example, the resistivity log constrains the sand fraction much more tightly than the gamma-ray log alone would do. Note that the standard deviations of sand fraction and HPT are only about 12–13% of their respective mean values, indicating that these key parameters are well constrained despite the high shale fraction and the uncertainty in water saturation.

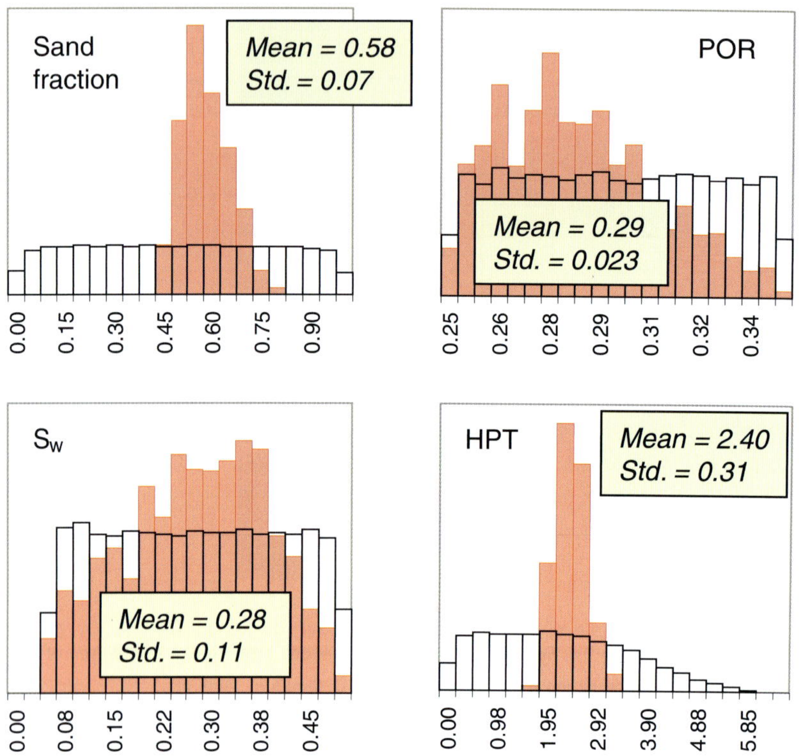

Figure 2.5. ***A priori*** **and inverted distributions of sand properties.** ***A priori*** **distributions are represented by empty histogram bars; inverted distributions are represented by shaded bars. Mean and standard deviation are posted for each inverted distribution.**

Field example

The deep-water sandstone example shown in Figure 2.6 focuses on a 14.5-m [47.6-ft] shaly interval between two thick channel sands. If this shaly interval had not been cored, its production potential might never have been recognized. Analysis of ultraviolet core photographs shows 65% net-to-gross, and core plug measurements of porosity and water saturation yield 1.5 m [4.9 ft] of HPT in this interval. On the other hand, conventional log analysis indicates only 15% net-to-gross and 0.3 m [1 ft] of HPT.

Application of the statistical Volumetric Laminated Sand Analysis (VLSA) technique to this 14.5-m-thick [47.6-ft-thick] interval is illustrated in Figure 2.7. Here the log averages are used as input to the VLSA inversion procedure. Note the significant improvement in HPT accuracy compared to the standard log analysis results: Net-to-gross of 55% instead of 15%, and HPT of 1.5 m [4.9 ft] instead of 0.3 m [1 ft]. The VLSA assessment of HPT agrees with the core-derived HPT and represents a 400% increase over the standard log evaluation estimate.

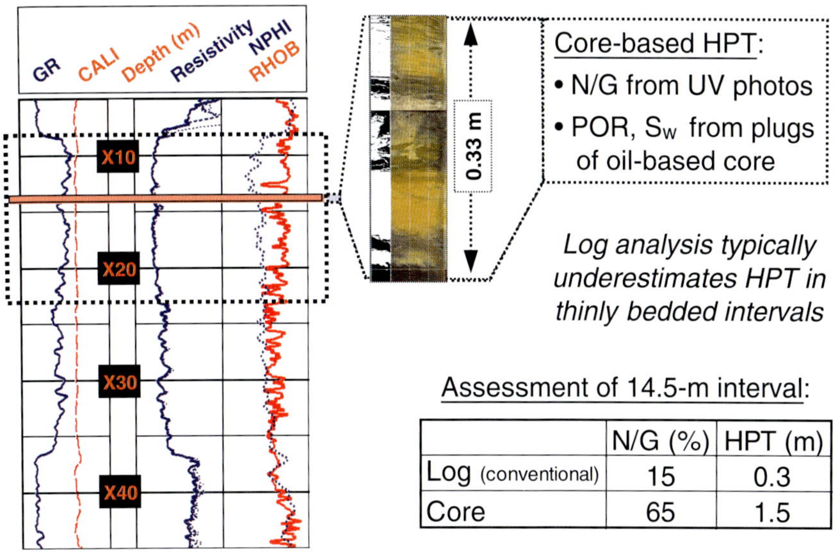

	N/G (%)	HPT (m)
Log (conventional)	15	0.3
Core	65	1.5

Figure 2.6. Log and core example of thinly bedded pay between two massive sandstones. Note mismatch of HPT from core and conventional log analysis methods for the 14.5-m [47.5-ft] interval outlined on the figure.

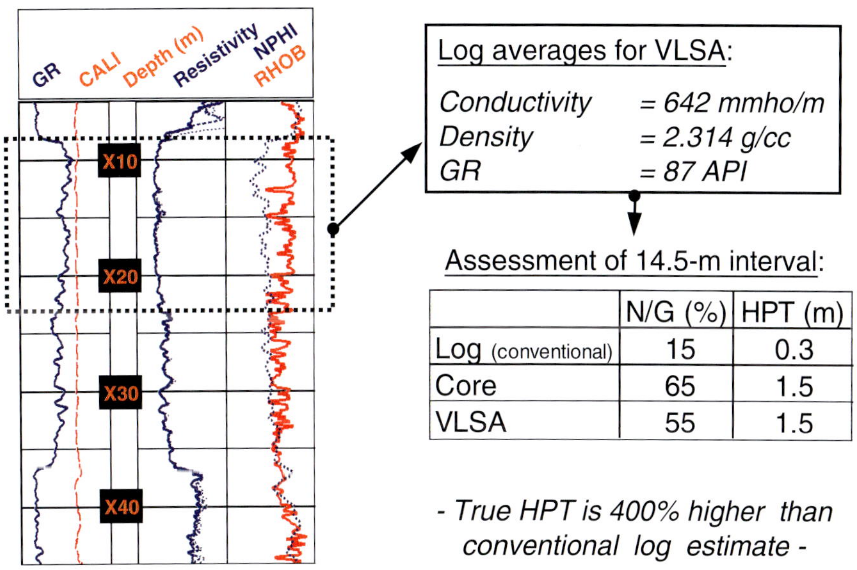

	N/G (%)	HPT (m)
Log (conventional)	15	0.3
Core	65	1.5
VLSA	55	1.5

- True HPT is 400% higher than conventional log estimate -

Figure 2.7. Same log example as Figure 7, but the low-resolution Volumetric Laminated Sand Analysis (VLSA) technique was applied. Note the good agreement between core- and VLSA-derived net-to-gross and hydrocarbon pore thickness (HPT) for this 14.5-m [47.5-ft] interval.

Conclusion

The strength of the Monte Carlo inversion technique used in VLSA is that all the logs simultaneously constrain all the parameters via the forward-model equations. This is in contrast to most of the published low-resolution approaches, where sand resistivity is calculated on a depth-by-depth basis from the measured resistivity via a shale-fraction correction derived from other logs [e.g., Bateman, 1990; Fylling, 1991; Van den Berg et al., 1996]. Such approaches tend to be quite sensitive to errors in the shale fraction.

A key element in this low-resolution technique is the set of mathematical models used to predict interval-average log response from the statistical earth model. For most logs a simple spatial average is adequate. For the resistivity (conductivity) log, the complication introduced by macroscopic anisotropy is modeled using a published equation [Chapter 10]. An additional complication resulting from finite lateral bed dimensions is modeled using effective medium theory [Chapter 6].

Summary

Two complementary methods have been developed to improve the estimation of hydrocarbon pore-thickness (HPT) in thinly bedded siliciclastic reservoirs. The first, generically termed the *high-resolution* method, is appropriate for reservoirs where the average thickness of sandstone and shale beds is at least 1 ft [.3 m]. The second, generically termed the *low-resolution* method, is applicable to any bed thickness and is most useful where bed thicknesses are less than 1 ft [.3 m]. A specific low-resolution technique introduced here is called Volumetric Laminated Sand Analysis (VLSA).

Both methods employ log-response models that account explicitly for the effects of thin beds on well-log measurements. Thus each method, within its proper domain of application, can produce estimates of HPT that are much improved when compared to conventional log analysis. The characteristics of the two methods are contrasted in Table 2.2.

Table 2.2. Characteristics of high- and low-resolution techniques for evaluating thin-bedded clastic reservoirs.

Characteristic	High-resolution method	Low-resolution method (VLSA)
Log requirements	Requires a log that can resolve all bed boundaries	No high-resolution log required
Bed definition and earth model	Detailed earth model; all bed boundaries and properties of individual beds are defined explicitly	Statistical earth model; individual beds not defined; a thin-bedded interval is characterized by proportions of defined *bed types*
Log response models	Compute explicit log responses at each depth from detailed earth model	Compute *interval-average* log responses from statistical earth model
Solution method	Iterative inversion of detailed log-response models	Monte Carlo inversion of interval-average log-response models
Form of solution	Explicit high-resolution depth-by-depth solution for lithology, porosity, and water saturation	Probabilistic solution for interval-average sandstone fraction, porosity, water saturation, and HPT
Treatment of uncertainty	Not explicitly treated; can produce multiple non-unique solutions	Explicit treatment of uncertainty via probabilistic solution

Fluvial Point Bar,
Upper Cretaceous,
Rock Springs Formation:
Superior, Wyoming, U.S.A.

Photo by Kevin Bohacs

3 Definitions and Geologic Occurrence of Thin Beds in Clastics

Overview

Introduction

This chapter discusses the petrophysical and geological usages of the term *thin bed* and introduces the definitions for *thin beds* and *very thin beds* that are used in the present publication. Thin beds occur in most siliciclastic depositional environments, and specific sub-environments where thin beds regularly occur are highlighted.

Contents

DOI: 10.1306/1157786A13220

Geological and Petrophysical Thin Beds

Thin beds

The term *thin bed* is used differently in different contexts. In particular, *thin bed* often has different meanings in the geological and the petrophysical literature. In this section we discuss these varying meanings and introduce the definition that we use in this publication.

Geological definitions

In the geological literature, *beds* have been defined as the most readily recognized layers of a sedimentary body, and are considered to be the basic building blocks of these bodies. Beds are bounded by depositional surfaces termed *bedding surfaces*; each surface is practically synchronous, and a bed can be considered an informal time-stratigraphic unit of limited areal extent representing a relatively short time span. From smallest to largest, the component layers of a sedimentary body are laminae, lamina-sets, beds, and bed-sets [Campbell, 1967]. Table 3.1 provides a more recent discussion of these terms [after Van Wagoner et al., 1990].

Table 3.1. Definitions of geologic layers.

Layer type	Definition	Time for deposition	Typical thicknesses
Lamina	Laminae are defined as the smallest megascopic geological layers. Laminae are uniform in composition and texture, and are never internally layered.	Min to hr	Less than 1 in. [2.5 cm]
Lamina-set	A relatively conformable succession of genetically related laminae bounded by surfaces of erosion, non-deposition, or their correlative conformities. Lamina-sets consist of a group or set of conformable laminae that compose distinctive structures in a bed.	Min to days	Less than 5 ft [1.5 m]
Bed	A relatively conformable succession of genetically related laminae or lamina-sets bounded by surfaces of erosion, non-deposition, or their correlative conformities. Not all beds contain lamina sets.	Min to yr	Inches to tens of ft
Bed-set	A relatively conformable succession of genetically related beds bounded by surfaces of erosion, non-deposition, or their correlative conformities. Beds above and below bed-sets always differ in composition, texture, or sedimentary structure from those composing the bed-set.	Longer period of time than beds	1 to 50 ft [.3 to 15 m]

In the geological literature, *thin beds* have been defined differently by different workers. One definition includes sedimentary beds ranging from 5 to 60 cm [approximately 2 in. to 2 ft] in thickness [Bates and Jackson, 1984]. Campbell [1967] limits *thin beds* to beds with thicknesses between 3 and 10 cm [approximately 1 to 4 in.]. Additional definitions for bed and lamina thicknesses are provided in Figure 3.1.

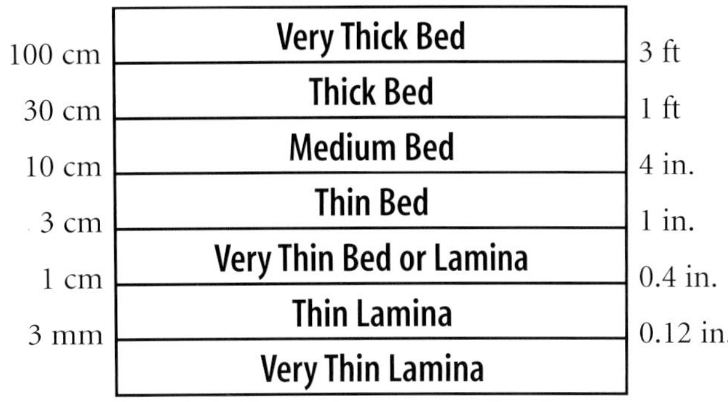

Figure 3.1. Terminology for bed and lamina thickness as defined by Campbell [1967].

Petrophysical beds

In formation evaluation, petrophysical, or log analysis applications, the term *bed* has been used to refer to intervals with consistent reservoir properties (e.g., clay volume, porosity, water saturation) based on log and/or core analysis techniques. The term *thin bed* has been used in a qualitative sense to refer to units of consistent petrophysical properties that are thinner than the vertical resolution of conventional logging tools. Recall from Chapter 1 that the vertical resolution of a logging tool is the thinnest bed in which a true reading can be obtained.

We define a *petrophysical bed* as

> any contiguous unit of rock with a narrow distribution of petrophysical characteristics that is bounded above and below by units with significantly different petrophysical characteristics.

For this definition, petrophysical characteristics include porosity, grain density, permeability, and capillary pressure behavior. Note that this definition could describe a geological bed, bed-set, lamina-set, or lamina.

Petrophysical thin beds

A *petrophysical thin bed* is

> a petrophysical bed with thickness between 1 in. [2.5 cm] and 2 ft [.6 m].

The 2-ft thickness is selected as a thin-bed cutoff because it is approximately equal to the vertical resolution of the porosity logs and highest-resolution resistivity logs (that is, these logs resolve the petrophysical bed boundaries and give a true measure of the petrophysical property near the center of the bed).

Petrophysical very thin beds

A *petrophysical very thin bed* is

> a petrophysical bed with thickness less than 1 in. [2.5 cm].

Petrophysical *very thin beds*, so defined, are near or below the resolution of image logs and conventional core plugs.

For the sake of brevity, the modifier *petrophysical* may be dropped in references to *thin beds* and *very thin beds* in the remainder of this publication.

Shaly Sands and Low-resistivity Pay

Shaly sands

A *shaly sand* may be defined as

> a sandstone in which quartz is the primary mineral, but clay and other associated minerals may be present in varying amounts, distributions, and particle sizes.

When clay minerals are present in a sandstone, the type, volume, and distribution of the clay will affect the well log response to that sandstone. In some cases, the effects of clay can significantly impact the analysis of hydrocarbons-in-place. Worthington [1985] gives an overview of methods for addressing these issues with an emphasis on dispersed clay, where bed properties are resolved by conventional logs.

The sandstone beds in thinly bedded siliciclastic reservoirs may consist of clean (nearly clay-free) or shaly (clay-rich) sandstones. The presence of clay within the sandstone may necessitate the application of shaly sand models as described by Worthington [1985]. However, the application of such models in thin beds is different than it is for thick beds (see Chapters 11–13). *Direct application of dispersed-clay shaly sand models in thinly bedded reservoirs will usually result in underestimation of hydrocarbon pore-thickness.* See Chapter 1 for illustrations.

Low-resistivity and bypassed pay

Because of the low vertical resolution of resistivity logs, thinly bedded, hydrocarbon-bearing intervals often yield a lower resistivity response than would be obtained from a thick, hydrocarbon-bearing sandstone. If these intervals are recognized, perforated, and hydrocarbon-productive, they may be referred to as *low-resistivity pay*. If these intervals are not recognized, they may end up as *bypassed pay*.

Note that thick, shaly sandstones can also be referred to as low-resistivity pay.

Note:

At the 1997 Annual AAPG convention, a session was devoted to the recognition and evaluation of "Low-Resistivity/Low-Contrast Pays and the Potential for Bypassed Pays." In most presentations, the occurrence of thin beds was cited as a primary cause of the "low-resistivity pay" in the Gulf of Mexico, Rocky Mountains, mid-continent United States, and Indonesia [Sneider and Kulha, 1997; Mullarkey and Dolly, 1997; Eubanks and Oraby, 1997; Frodesen and Lawrence, 1997].

Geologic Occurrence of Thinly Bedded Sandstone Reservoirs

Thin beds are ubiquitous

Thin beds occur in all siliciclastic depositional systems, with some environments consisting largely of thin beds. Table 3.2 provides an overview (not exhaustive) of the depositional environments where thin beds are more or less likely to occur.

Table 3.2. Summary of Depositional Environments.

Depositional System	Thin-bed-prone	Not thin-bed-prone
Deep-water	Overbank/levee deposits Distributary lobe Channel margin Hemipelagic	Channel axis Debrites (sandy or muddy)
Beach/shoreface	Lower shoreface Distal lower shoreface	Foreshore Upper shoreface
Deltaic	Delta front Prodelta	Stream-mouth bar
Tidal/estuarine	Sandy tidal channel Intertidal sand flats	Subtidal
Fluvial	Point bars (meandering stream) Levees Terminal splay (overbank)	Braided streams Channel sands Channel lag deposits Fluvial bars Alluvial fans
Aeolian	Interdune Wind-rippled deposits	Cross-bedded dunes

Figures 3.2 through 3.8 show idealized vertical successions for these deposits and highlight the positions within the intervals characterized by thin beds.

Deep-water submarine-fan environment

The sandstones in Figure 3.2 represent a sharp-based, fining-upward succession of thin-bed sandstones and interstratified mudstones in a deep-water Gulf of Mexico reservoir. These sediments were deposited primarily by low-density turbidites in unchannelized distributary lobes. These unchannelized distributary lobes have high net-to-gross (72–87%) and excellent lateral continuity.

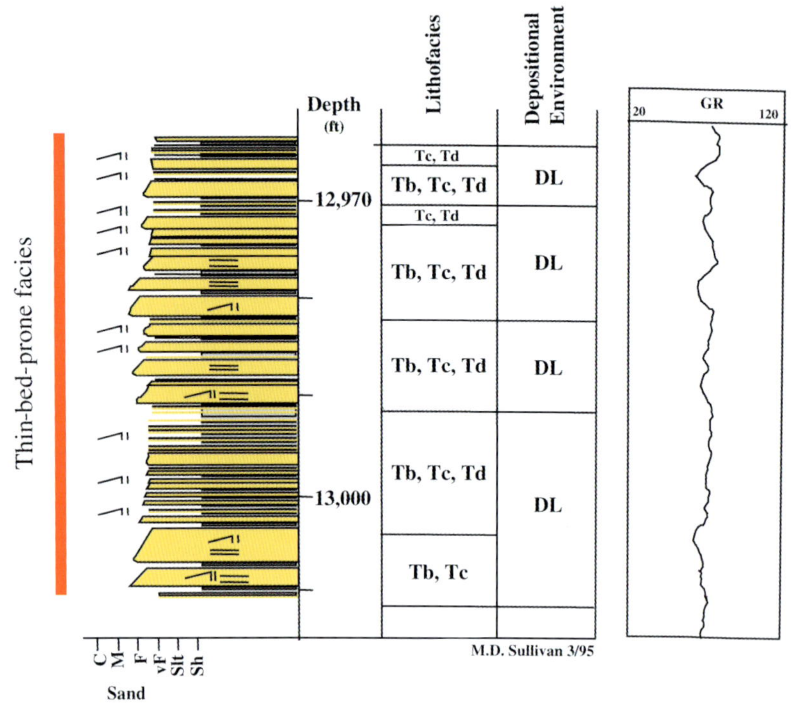

Figure 3.2. Stratal characteristics of deep-water distributary lobes [Original courtesy of M. D. Sullivan and D. Bagley].

Lithofacies Associations

T a = massive sandstone
T b = planar-stratified sandstone
T c = current ripple-stratified sandstone
T d = planar-stratified shaly sandstone
T e = planar-stratified to massive mudstone

The sandstones in Figure 3.3 represent channel-axis and channel-margin deposits in another deep-water Gulf of Mexico reservoir. The channel-axis deposits are dominated by amalgamated, thick-bedded, massive sandstones (Ta lithofacies) deposited by high-density turbidites. The channel-margin deposits are characterized by heterolithic mixtures of planar-stratified mudstones and thin-bedded low-density turbiditic sandstones.

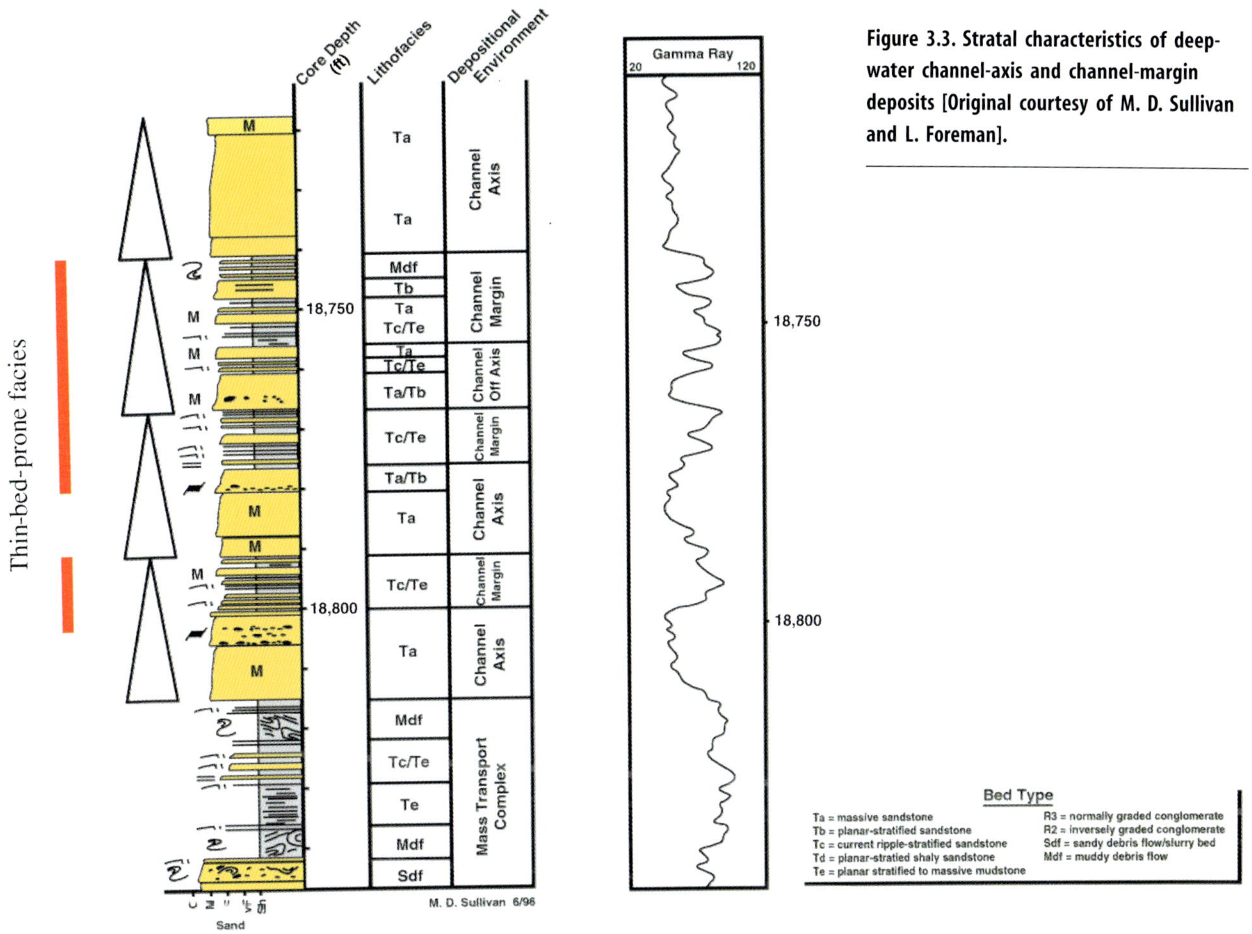

Figure 3.3. Stratal characteristics of deep-water channel-axis and channel-margin deposits [Original courtesy of M. D. Sullivan and L. Foreman].

Beach environment

The parasequence in Figure 3.4 forms in a beach environment on a sandy, wavy- or fluvial-dominated shoreline.

The red bold line indicates the units where thin sandstone beds often occur, primarily restricted to the proximal-to-distal lower shoreface.

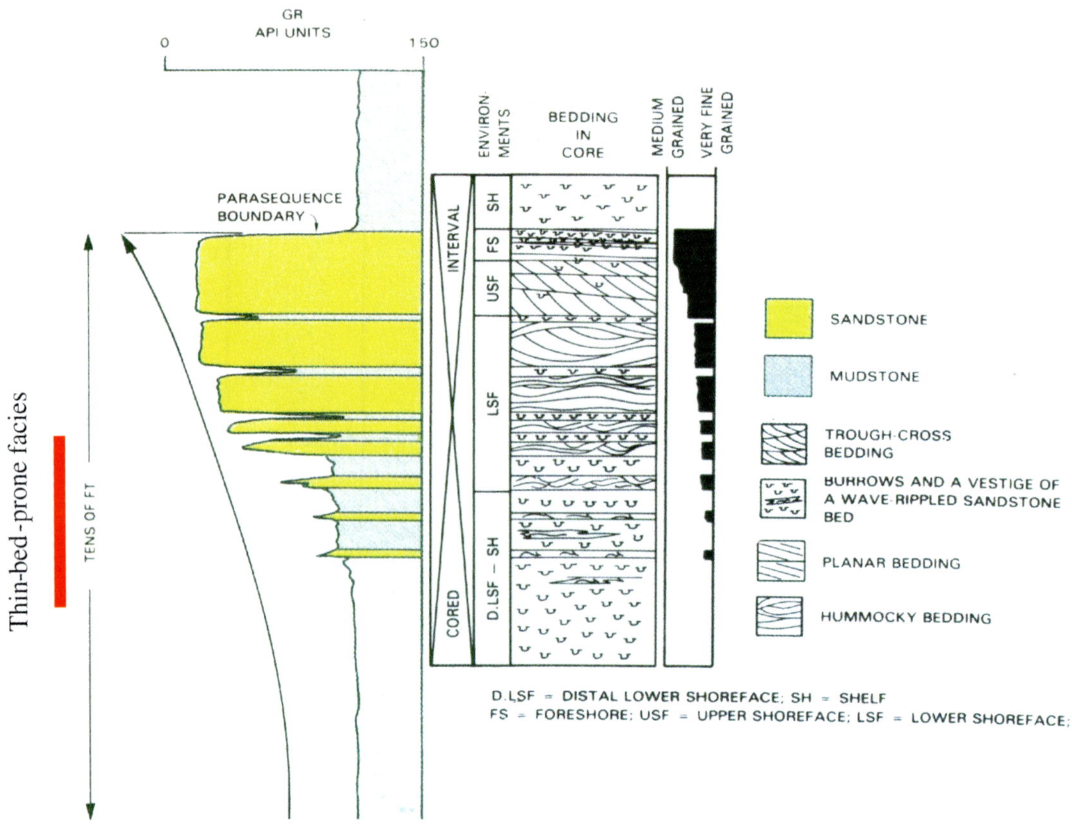

Figure 3.4. Stratal characteristics of an upward-coarsening beach parasequence [Van Wagoner et al., 1990].

Delta environment

The parasequence in Figure 3.5 forms in a deltaic environment on a sandy, fluvial-, or wave-dominated shoreline.

The red bold line indicates the units where thin sandstone beds often occur; multiple stacked thin sandstone beds occur throughout the delta front (DF), and prodelta (PRO D) deposits.

Note that the thick sand unit at the top represents outer stream-mouth bar (OSBM) deposits, which are internally composed of burrowed and trough-cross-bedded strata.

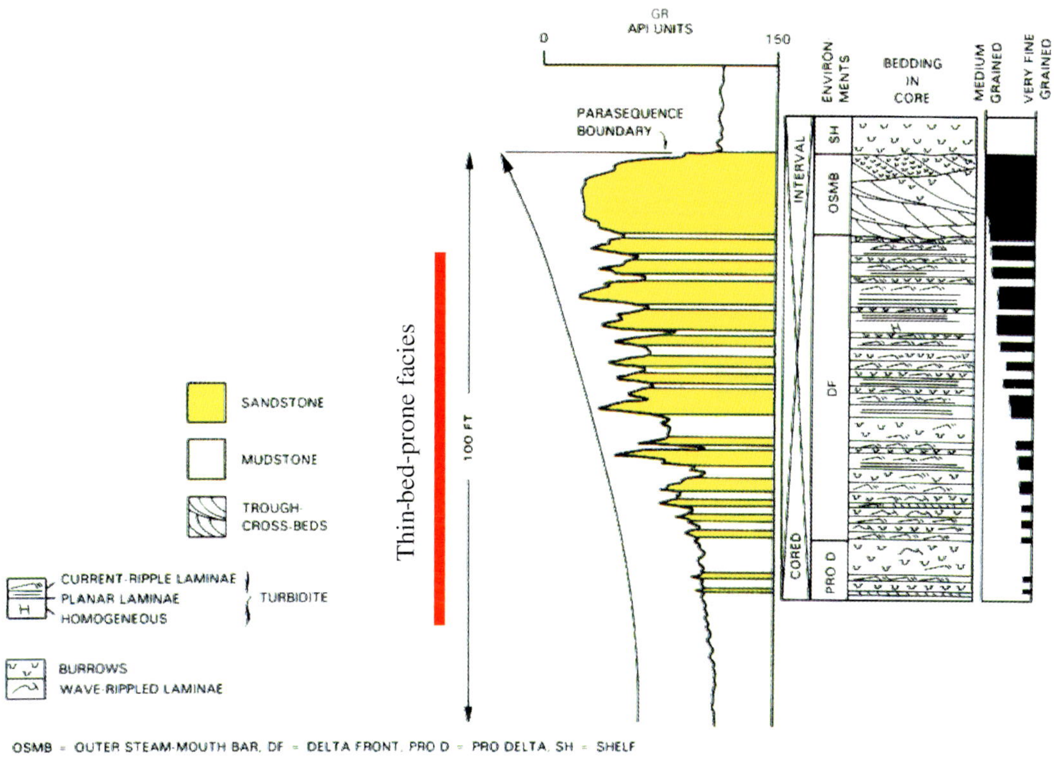

Figure 3.5. Stratal characteristics of an upward-coarsening deltaic parasequence [Van Wagoner et al., 1990].

Tidal/estuarine environment

The types of parasequences in Figure 3.6 form in a tidal flat to subtidal environment on a muddy, tide-dominated shoreline.

Thin beds commonly occur in the intertidal (INT) environments (where the bedding types include flaser, lenticular, and wavy beds).

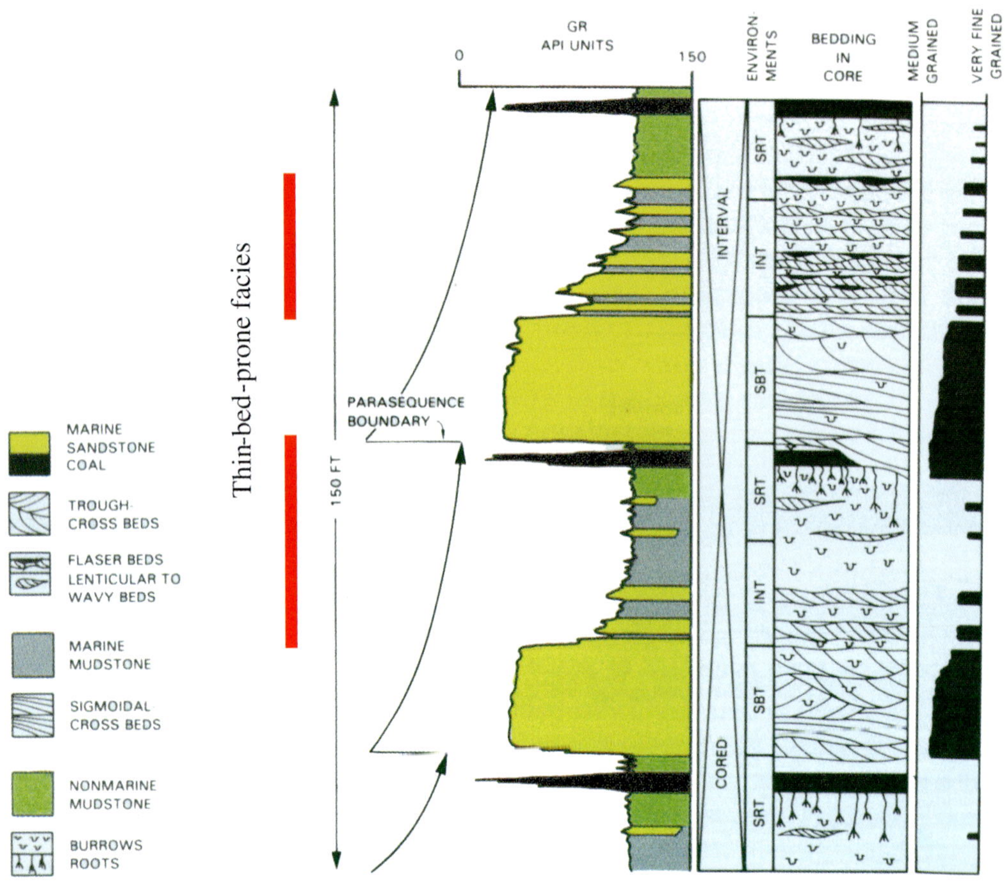

Figure 3.6. Stratal characteristics of two upward-fining tidal/estuarine parasequences [Van Wagoner et al., 1990].

Fluvial (braided-stream) environment

The sandstones in Figure 3.7 represent braided-stream (fluvial) deposits. The stacked bars make up a "blocky" profile on wireline logs, with very little mud occuring within the sand body. The thinly bedded intervals tend to be associated with current-rippled cross-laminated terminal splays located at the top of the sequence.

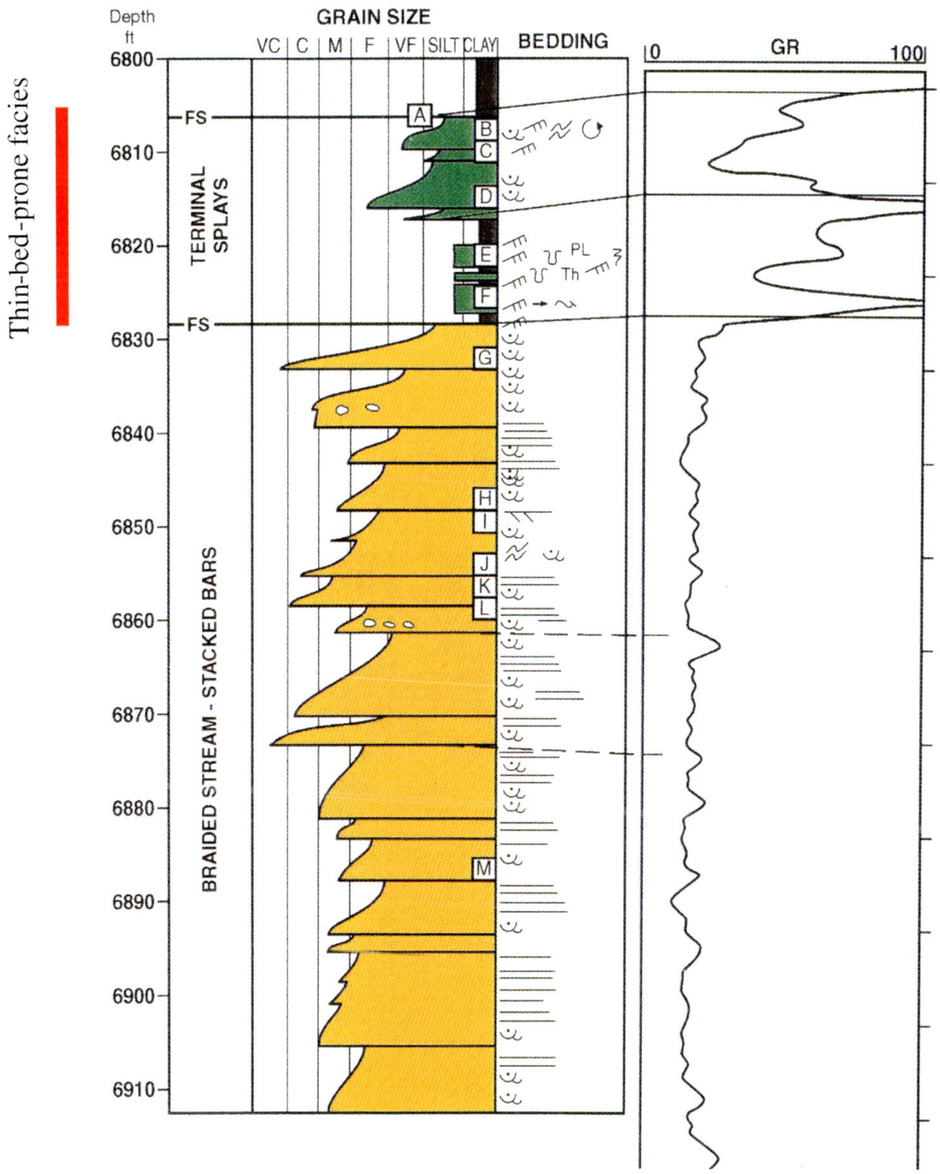

Figure 3.7. Stratal characteristics of stacked braided-stream deposits (core description by J. C. Van Wagoner and C. R. Jones).

Fluvial (point-bar) environment

The sandstones in Figure 3.8 represent stacked point-bar (fluvial) deposits. Overall a fining-upward grain-size pattern is observed, with an increase in mud content and thin beds upsection. Thinly bedded intervals tend to be associated with upper point bars, crevasse splays, and flood plain deposits.

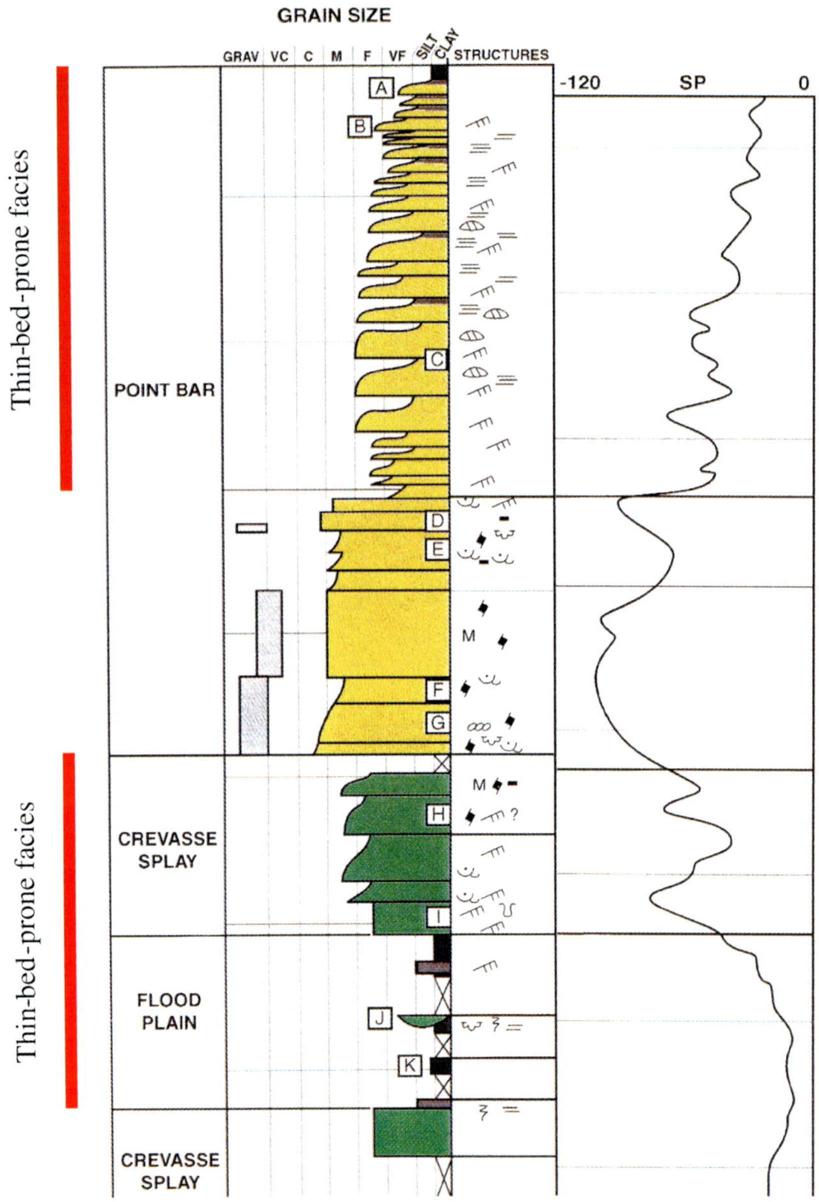

Figure 3.8. Stratal characteristics of stacked point-bar deposits (core description by S. J. Friedman).

Summary

The terms *bed* and *thin bed* are often used differently in the geological and the petrophysical literature. For the purposes of this publication, we define a *petrophysical bed* as any contiguous unit of rock with a narrow distribution of petrophysical characteristics that is bounded above and below by units with significantly different petrophysical characteristics.

For our purposes, a *thin bed* is a petrophysical bed with thickness between 1 in. [2.5 cm] and 2 ft [.6 m]. The 2-ft [.6-m] thickness is set as the thin-bed cutoff because it is approximately equal to the vertical resolution of the porosity logs and highest-resolution resistivity logs. A *very thin bed* is a petrophysical bed with thickness less than one inch. Very thin beds, so defined, are below the resolution of image logs and conventional core plugs.

Thin beds occur in all siliciclastic depositional systems, with some environments consisting largely of thin beds. Table 3.2, repeated here, provides an overview (non-exhaustive) of the depositional environments where thin beds are more or less likely to occur.

Table 3.2. Summary of Depositional Environments.

Depositional System	Thin-bed-prone	Not thin-bed-prone
Deep-water	Overbank/levee deposits Distributary lobe Channel margin Hemipelagic	Channel axis Debrites (sandy or muddy)
Beach/shoreface	Lower shoreface Distal lower shoreface	Foreshore Upper shoreface
Deltaic	Delta front Prodelta	Stream-mouth bar
Tidal/estuarine	Sandy tidal channel Intertidal sand flats	Subtidal
Fluvial	Point bars (meandering stream) Levees Terminal splay (overbank)	Braided streams Channel sands Channel lag deposits Fluvial bars Alluvial fans
Aeolian	Interdune Wind-rippled deposits	Cross-bedded dunes

Deep-water (Distal Levee),
Permian,
Laingsburg Formation:
Paarclekraal, Laingsburg, South Africa

Photo by Anthony Sprague

4 Identification of Thin Beds using Well Data

Overview

Introduction

Because of their limited vertical resolution, standard logging tools generally do not allow the direct identification of thin beds or the direct measurement of the physical properties of individual thin beds. A logging tool records a signal that is a mixture of the formation properties within the tool's vertical range of investigation. This vertical range is typically larger than the thickness of an individual thin bed, and varies for different tools.

Given these limitations, this chapter discusses several methods for identifying or inferring the presence of thinly bedded formations using well logs and other wellbore data.

Contents

DOI: 10.1306/1157787A13220

Thin Beds and Tool Resolution

Definition: Vertical resolution

As discussed in Chapter 1, the *vertical resolution* of a logging tool is defined as the thickness of the thinnest bed in which a true reading can be obtained.

Vertical resolution and bed thickness

Figure 4.1 shows a comparison of bed thickness and typical resolution of logging tools, core plugs, and thin sections. Refer to Chapter 3 for a discussion of the geological and petrophysical definitions of thin beds.

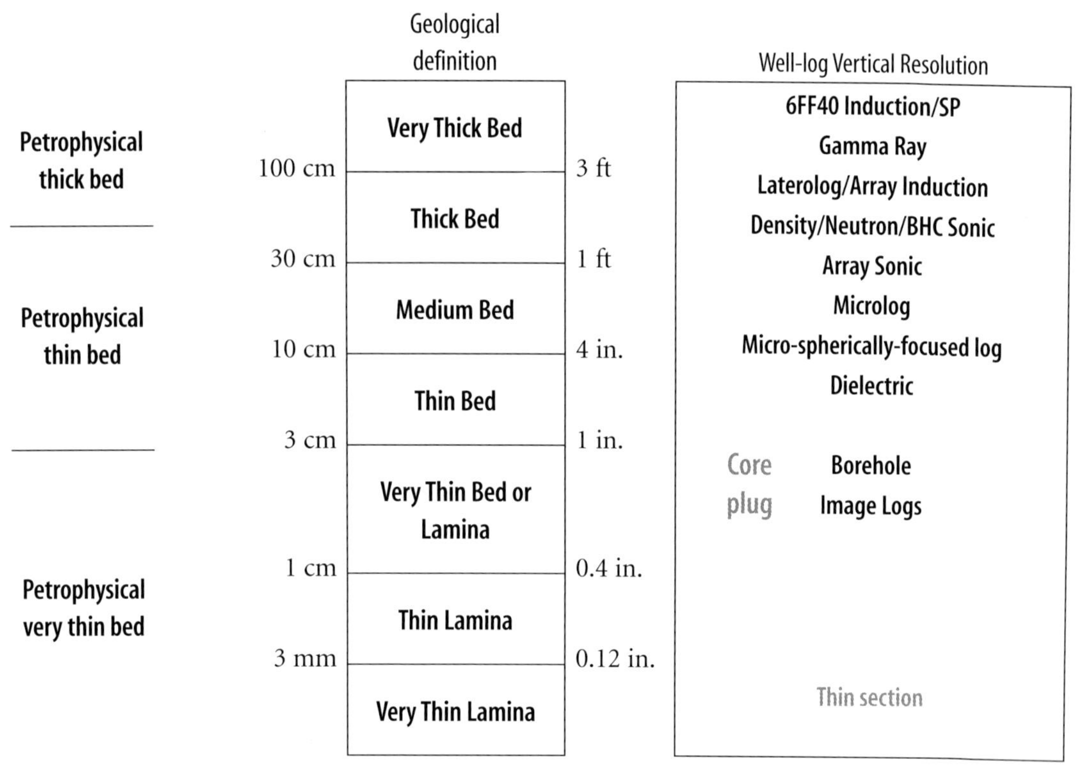

Figure 4.1. Comparison of petrophysical beds, geological beds [after Campbell, 1967], and well-log vertical resolution.

Definition: Detection limit

Loosely defined, the *detection limit* of a logging tool is the thickness below which a bed cannot be distinguished from its neighboring shoulder beds. This limit is not a fixed property of the logging tool, but also depends on the contrast between the bed and its neighbors. Most thin-bedded reservoirs contain beds that are below the detection limits of conventional logs. In particular, *very thin beds* are usually below the detection limits of all the logs used in conventional formation evaluation. This is the factor that can render difficult, or even impossible, the unambiguous identification of a thin-bedded reservoir based on standard logs alone.

Are Thin Beds Present?

Direct and indirect indicators of thin beds

In new exploration and production wells, the petrophysicist may elect to design a logging and coring program that will address the thin bed issue explicitly [Chapter 5]. In existing wells, however, one must work with existing data to identify thin-bedded reservoirs and evaluate their potential for production. Several kinds of data may alert the analyst to the presence of thin beds. These data types can be categorized as *direct* and *indirect* indicators.

- *Direct* indicators of thin beds include:
    - Conventional core and core images
    - Sidewall cores
- *Indirect* indicators of the presence of thin beds include:
    - Borehole image logs (electrical and acoustic)
    - Dipmeter logs
    - Conventional and high-resolution logs
    - NMR logs
    - Multi-component resistivity logs
    - Mud logs and drilling data

In addition, production-related data may give further indirect evidence of the presence of thin bedding:

- Wireline formation tests
- Historical production data

Each of these indicators is discussed below.

Direct Indicators of Thin Beds

Conventional core and core images

For establishing thin-bed ground-truth in an interval, there is no substitute for conventional core. Moreover, digital core images can provide a convenient means of comparing core to log response, and determining the volumes of sand and shale with a high confidence level [Chapter 8]. Figure 4.2 shows a 20-ft [6-m] slabbed core interval depth-aligned to an electrical borehole image log (EBI*).

Both the white- and ultraviolet-light core images can be used to differentiate oil-bearing sandstone (light) from shale (dark) layers. In the example in Figure 4.2, laminations as thin as a few millimeters are readily discernible.

**Note: We use a non-standard name and acronym (electrical borehole image, EBI) to refer generically to anyone of a collection of micro-electrical borehole imaging tools. For reference to specific tools, see Chapter 9.*

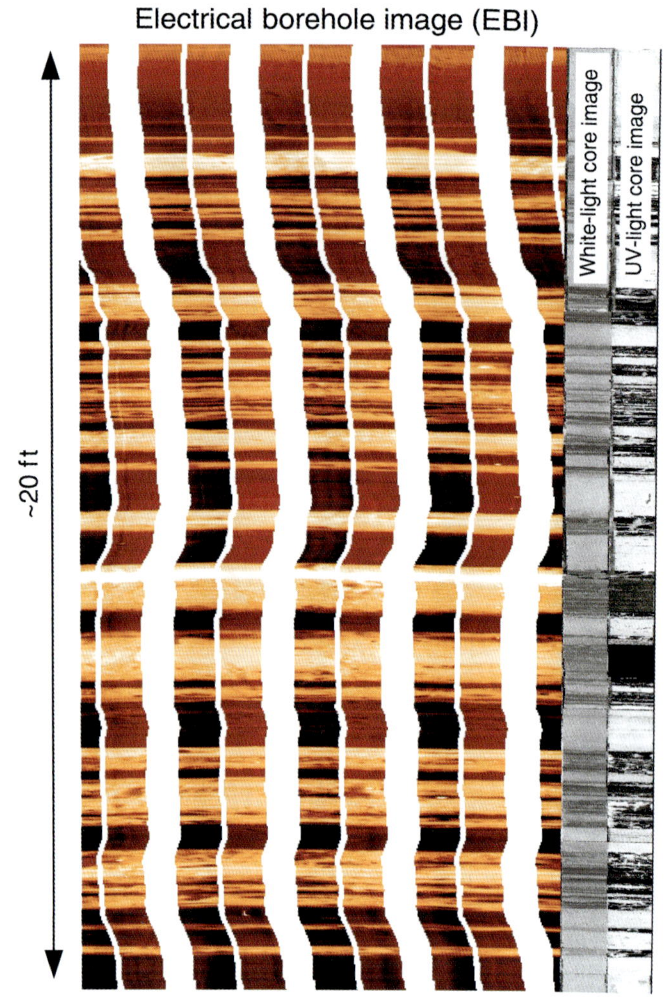

Figure 4.2. Borehole image log (left) correlated with core images (right)

Sidewall cores

If conventional core is unavailable, then percussion or rotary sidewall cores may be the only direct evidence of thin beds. Of course, one must consider the probability that both thin sandstone and shale beds would be encountered given the frequency of the sidewall data.

Indirect Indicators of Thin Beds

Borehole images

Digital electrical and acoustic borehole images [Chapter 9] provide unique log data that is, in many respects, similar to conventional core images. Like core images, they provide a high-resolution view of the formation that allows high-resolution geological and petrophysical interpretations. Assuming high-quality data, accurate calibration, and proper processing, the high-resolution images can provide:

- Identification of thin-bed occurrence;
- Quantification of bed thickness;
- Interpretation of gross sandstone thickness;
- Identification of descriptive rock facies; and
- In some cases, interpretation of hydrocarbon-bearing sandstone thickness.

In Figure 4.2, the high-quality electrical borehole image (EBI) data detect beds as thin as 2 cm [.8 in.] or less. Note that the shale beds, which appear dark in the core images, are bright (or more resistive) in the EBI image even though the sandstones are hydrocarbon bearing. This is opposite the usual pattern and suggests that the hydrocarbons have been flushed from the near-wellbore region that is investigated by the tool.

Borehole images are not equivalent to core images. One must exercise caution when interpreting lithology solely from borehole images. Electrical borehole images represent variations in electrical conductivity; acoustic borehole images represent variations in acoustical rock properties. These attributes are related to lithology, but can be affected by many other factors such as fluid type and chemistry, mud properties, speed-correction errors, pad contact or hole rugosity. Figure 4.3 illustrates the difficulty in interpreting lithology solely from an electrical borehole image log where both hydrocarbon- and water-bearing sandstones are present.

At first inspection, the upper part of the EBI image in Figure 4.3 might be interpreted as containing thinly interbedded shales (dark) and sandstones (bright). Closer inspection indicates a strong cor-

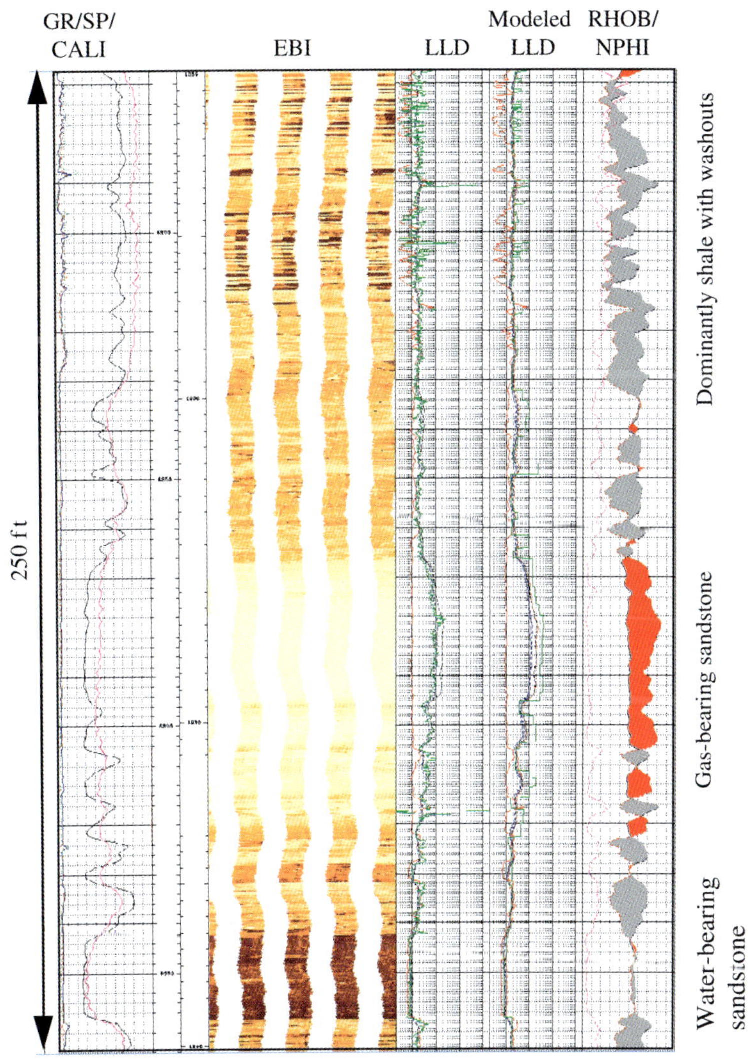

Figure 4.3. Integrated borehole image and log interpretation for a deep-water turbidite reservoir.

relation between the caliper and the dark intervals, suggesting that the EBI is reading mostly drilling mud in the washouts. Moreover, the light intervals in the upper section are essentially the same intensity as the EBI response in deeper thick shale sections. Thus, this upper interval is interpreted as primarily shale, consistent with the gamma-ray and neutron/density response, and the lithology of the washed-out intervals is not known.

The gas-bearing sandstone, just below the middle of the interval, appears bright on the EBI image because of the high resistivity (low conductivity) in the near-wellbore region. Note, however, that the sandstone at the bottom of the interval is water filled, and appears dark on the EBI image.

Although the presence of thin beds may be inferred from borehole images, it is often impossible to interpret lithology from these images alone.

Dipmeter logs

The vertical resolution of standard dipmeter logs is similar to that of electrical borehole-imaging tools. For most dipmeter tools a maximum of eight conductivity traces is available, versus up to 192 for EBI logs. Although the occurrence of thin beds can be inferred by examination of the activity on each trace, thin beds are best inferred from dipmeter logs by high correlatable activity on raw dipmeter resistivity curves, and generally consistent dip magnitude and azimuth.

In the example in Figure 4.4, a 6-m [19.7-ft] section of interpreted dipmeter data is shown. In the left track, the tadpoles indicate numerous high-quality and consistent dips. In the right track, the correlations associated with the tadpoles are presented. Note that it is reasonable to infer from these data that this interval is characterized by thin, parallel beds. As with the EBI, however, one must be aware of tool limitations when attempting to determine lithology from dipmeter data alone.

Figure 4.4. Dipmeter plot (1 m [3.3 ft] per division). Correlated high-frequency features on resistivity curves suggest thin parallel beds [Serra and Andreani, 1991].

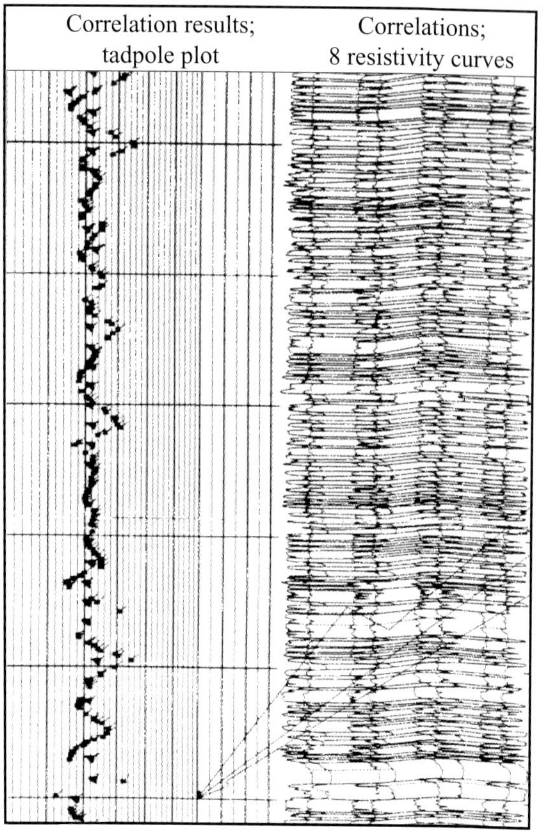

Conventional and high-resolution logs

Following is a list of several methods used to infer the presence of thin beds from standard and high-resolution log data [modified after Serra and Andreani, 1991]:

- Spontaneous potential (SP) and gamma ray (GR) indicate a shaly formation, often with high-frequency variations.
- Porosity logs exhibit high-frequency variations, and the density curve may mirror the neutron curve. Density and neutron are closer together compared to a massive shale. In some interbedded formations, the density curve will show variation when the neutron curve is relatively constant.
- If the thin sands are hydrocarbon-bearing, the resistivity curve is slightly elevated relative to surrounding shales.
- Spherically focused log (SFL) appears spiky, indicating that thin beds of high resistivity are not resolved by the deeper-reading induction tool.
- Increased separation between shallow and deep resistivity measurements relative to shales, indicating that permeable intervals are present.
- High-frequency amplitude variation on shallow resistivity and/or dielectric log(s), especially on the dielectric attenuation curve.
- Microlog separation indicating thin permeable beds.
- Caliper less than bit size, indicating mudcake in permeable zones.

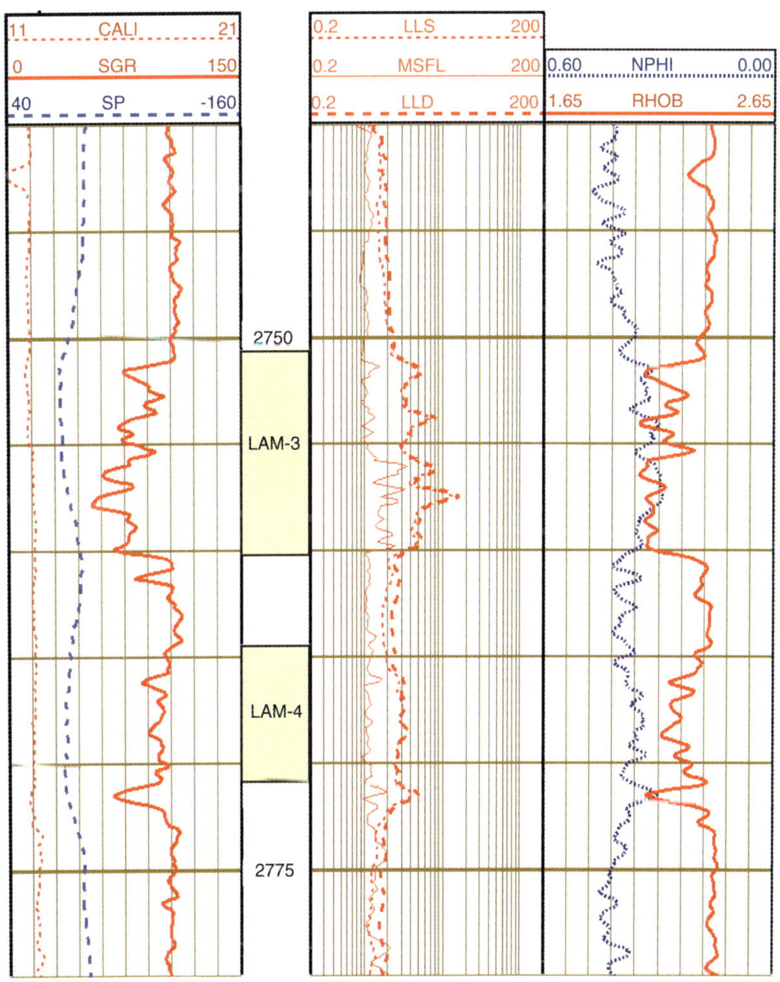

Figure 4.5. Conventional well-log response to laminated sandstone and shale intervals.

Figure 4.5 illustrates several of these points with standard well logs for two laminated sandstone intervals. For the upper laminated sand (LAM-3), both the SP and gamma-ray curves indicate the presence of sandstone, but the gamma-ray curve never measures the low GAPI values expected for clean sandstones. These observations, and the high frequency of the SGR curve, suggest that thin shale interbeds are present in this interval. The neutron/density curves indicate interbedded sandstones and shales, with the two curves mirroring each other. The shallow resistivity curve (MSFL) suggests thin hydrocarbon-bearing beds, with the overall resistivity response higher than surrounding thick shales.

For the laminated sand LAM-4, the SP indicates a sandy interval, but the gamma-ray curve gives some indication of thin sandstones interbedded in a primarily shale interval. Overall, the gamma-ray response is slightly lower than in the underlying shale (below 2773 m [9098 ft]). As with LAM-3, the neutron and density curves mirror each other, but the separation indicates a shalier interval than LAM-3. Also, the deep resistivity curve is slightly higher than in the surrounding thick shales, suggesting the presence of thin, hydrocarbon-bearing sandstones.

NMR logs

The NMR T_2 distribution [Ye and Rabiller, 2000] may often be used to infer the presence of a thinly bedded reservoir where conventional logs apparently see only shale. Figure 4.6 shows such an example. The data is from an offshore deepwater turbidite deposit that was fully cored [op. cit.].

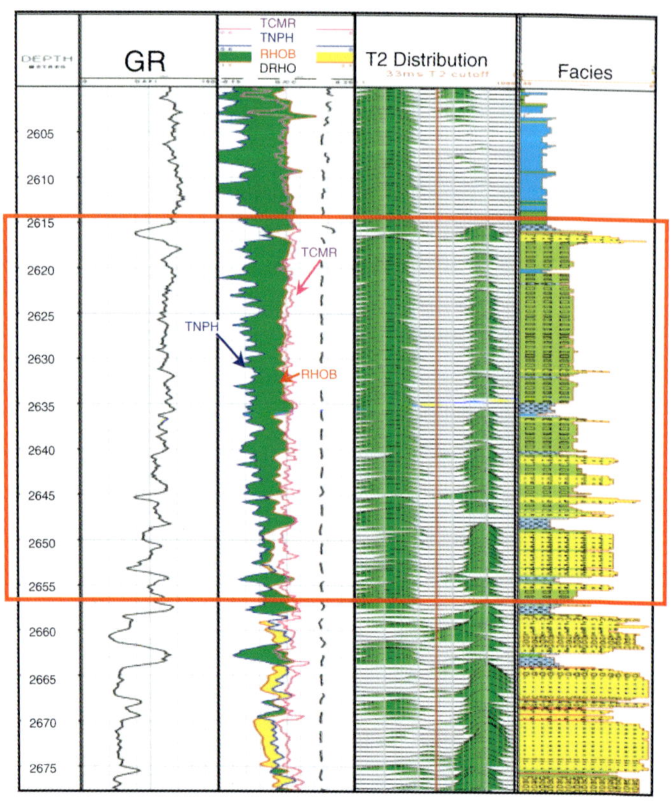

Figure 4.6. NMR T_2 distribution (Track 3) shows the presence of thinly interbedded sands and shales in the interval between 2615 and 2657 m [8580 and 8718 ft] (red box) [Ye and Rabiller, 2000].

Depth is in meters. Track 1 shows the GR log. Track 2 contains Schlumberger neutron porosity (TNPH), bulk density (RHOB), and CMR Total Porosity (TCMR) logs, along with density correction (DRHO). The shading between TNPH and RHOB is as follows: Green for shale and yellow for sand. With this information, one can readily identify the thick reservoir sands below 2657 m [8718 ft]. But what is happening at depths shallower than 2657 m [8718 ft]?

Track 3 shows the CMR T_2 distribution. In this distribution, amplitudes to the left of the red *cutoff* line represent bound water or shales. Amplitudes to the right of the cutoff indicate or suggest big pores or moveable fluids. Hence, the only likely interpretation is that between 2615 and 2657 m [8580 and 8718 ft] the formation consists of thin sands interbedded with shales. Shallower than 2615 m [8580 ft], there is only a thick shale. The T_2 distribution in this case provides information on reservoir quality, facies, and potential reservoir. Track 4 (a facies display) is included to confirm the information derived from NMR.

Multi-component induction logs

The multi-component induction logging tool can, in principle, provide a strong indicator for the presence of thinly bedded reservoirs. In the simplest scenario, where the borehole penetrates perpendicular to the horizontal bedding planes and the individual sand and shale beds are planar and isotropic, such a tool will measure two resistivities [Gianzero et al., 2002]. First is the horizontal resistivity (R_h), which is measured by currents parallel to the bedding planes and is equivalent to the standard induction-log resistivity. Second is the vertical resistivity (R_v), which is measured by currents transverse to the bedding planes. As discussed in Chapter 1, R_h is extremely sensitive to the low-resistivity interbedded shales, while R_v is much more sensitive to the high-resistivity hydrocarbon-bearing sands. Figure 4.7 is a synthetic log illustrating R_h and R_v in a case where the sand fraction is only 30%. The gamma ray, density and neutron porosities, and R_h barely hint at the presence of hydrocarbon sand in the interval from 9010 to 9030 ft [2746 to 2752 m], while the R_v measurement responds strongly to the resistive sand. Sand and shale beds in this example have the properties listed in Table 1.3.

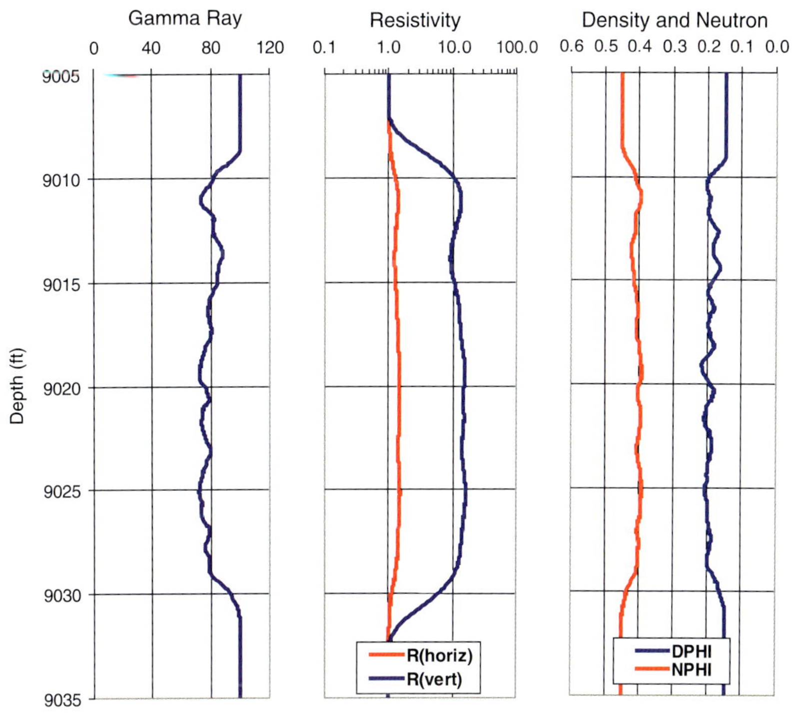

Figure 4.7. Synthetic triaxial resistivity measurements in a thinly bedded formation with V_{sh} = 70%.

Mud logs and drilling data

Mud log and other data collected while drilling can provide an indirect indication of thin hydrocarbon-bearing beds. Some potential clues for intervals that appear very shaly on gamma ray, resistivity, and density-neutron logs include:

- Mud gas or oil shows;
- Increased rate of penetration (ROP);
- Increased sandstone in the cuttings or cuttings description; and
- Lost returns.

Production Data

Wireline test

A wireline pressure tester can be employed to test for the presence of productive thin sandstone beds in intervals that appear shaly on standard logs.

With an inflatable packer module and one or more pump-out modules, new-generation wireline testing tools can be used to test prospective thin or very-thin bed intervals. With this tool, pressure-buildup data and fluid samples can be obtained from thinly bedded intervals where conventional probe testing would often fail, e.g., due to multiple seats on shale beds.

In Figure 4.8, the packers are shaded blue. Fluid enters the formation tester through the small orifice shown in red. If fluid in the formation flows radially into the borehole, then the data collected can be used to obtain an estimate of permeability-thickness in a region close to the borehole.

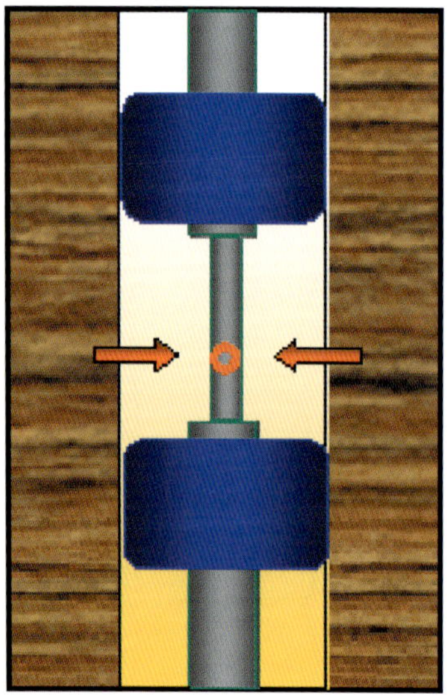

Figure 4.8. Wireline formation tester with dual inflatable packers.

Historical production data

Occasionally in a mature field, historical production data has alerted the producing company to the existence of production from previously unrecognized thin beds. For example, cumulative oil production may begin to approach or exceed the theoretical maximum production (MP) given by

$$MP = OOIP \times RE. \tag{4.1}$$

In Equation 4.1, OOIP is the original oil-in-place (Equation 1.1) and RE is the theoretical maximum *recovery efficiency* for the reservoir given by:

$$RE = [S_{oi} - S_{or}] \,/\, S_{oi} \tag{4.2}$$

In Equation 4.2, S_{oi} is the average initial oil saturation and S_{or} is the average residual oil saturation. Such an occurrence may lead the field petrophysicist to re-examine log and core data to look for previously unsuspected, productive thin-bedded intervals.

Summary

The *vertical resolution* of a logging tool is defined as the thickness of the thinnest bed in which a true log reading can be obtained. The *detection limit* of a logging tool is the thickness below which a bed cannot be distinguished from its neighboring shoulder beds. Very thin beds are usually below the detection limits of all the standard-resolution well logs. This factor often renders difficult, or even impossible, the unambiguous identification of a thin-bedded reservoir based on conventional logs alone.

The only unambiguous *direct* indicators for the presence of thin beds are cores and core photos. All well logs, including borehole image logs, are *indirect* indicators: They may provide evidence for the presence of thin beds, but such evidence is rarely unambiguous.

Borehole image logs provide the best indirect evidence for thin beds. These logs must be interpreted with due regard for environmental effects; but in the best case, they can be used to locate bed boundaries, build high-resolution earth models, and identify bed types. Ideally the interpretation of borehole image logs should be calibrated by comparison to cores or core images.

Standard logs may show several signs that provide evidence for the presence of a thin-bedded hydrocarbon reservoir. The gamma ray usually appears shaly, sometimes with high-frequency variations. Density and neutron porosity mirror each other with varying degrees of shale separation and spikiness. Resistivity may be elevated slightly relative to surrounding shales, but is subdued relative to thick hydrocarbon sands. There may be some separation between shallow and deep resistivity, indicating permeable beds. Higher-resolution logs, such as shallow resistivities, may show high-frequency variations.

The Nuclear Magnetic Resonance (NMR) free-fluid index, despite its low vertical resolution, can provide very strong evidence for the presence of productive thin-bedded reservoir in zones that appear predominantly shaly on conventional logs.

The new multi-component induction tools can provide estimates of both the vertical and the horizontal components of resistivity. Standard induction and focused-current tools measure only the horizontal component. The combination of both components is a much stronger indicator for a productive thin-bedded reservoir than the horizontal component alone.

Delta Front,
Cretaceous,
Panther Tongue Member of Star Point Formation:
Hardscrabble Canyon, Utah, U.S.A.

Photo by John Van Wagoner

5 Drilling, Coring, and Logging Programs for Thinly Bedded Formations

Overview

Introduction

This chapter presents some general considerations on the design of drilling, coring, and logging programs for thinly bedded clastic reservoirs.

As was discussed in Chapter 3, thin beds can occur in all clastic depositional systems. Therefore, for exploration and appraisal wells, it is recommended that log data be collected under the assumption that thin beds may be encountered. The continued requirement for such data can then be assessed in subsequent development wells.

Contents

DOI: 10.1306/1157788A13220

Drilling Considerations in Thinly Bedded Formations

Drilling mud

The acquisition of high-quality data for interpreting thin beds requires a good borehole. Appropriate drilling mud composition can make the difference between a "gun-barrel" borehole and one where shales are washed out and acquisition of high-quality data is impossible.

- Water-based muds should be formulated to avoid reactivity with formation shales.
- Often the best hole conditions can be obtained with oil-based or synthetic muds. Properly planned oil-based drilling also affords the possibility for collecting oil-based cores that can be used to measure in-situ water saturation.
- Electrical borehole image (EBI*) logs are affected by the type and composition of the drilling mud. Some EBI tools are designed to work in oil-based muds, and these tools tend to perform best when the mud's water fraction is relatively high. If the well is cored, a high water fraction in the mud may adversely affect the measurement of in-situ water saturation on core plugs.

Borehole size

Typically, logging tools that make direct contact with the borehole wall are designed for optimal contact in 8.5-in. [21.6-cm] diameter wellbores. The fixed curvature of the contact pads may cause some loss of contact in larger or smaller boreholes. Most high-resolution log measurements are made with these contact-type tools. Therefore, such high-resolution logs may tend to be of higher quality in an 8.5-in. [21.6-cm] diameter hole than in a 12.5-in. [31.8-cm] or 17-in. [43-cm] hole.

Coring Programs for Thinly Bedded Formations

Conventional coring

Core descriptions, core measurements, and core photographs are essential tools for the rigorous evaluation of thin-bedded reservoirs. Acquisition of cores and core photographs are discussed in Chapters 7 and 8, respectively. Following are a few general considerations on coring and core analysis in thinly bedded formations.

- A suite of high-quality core photographs, taken in both white light and ultraviolet light, is a key tool in the evaluation of a thin-bedded reservoir. Core photography is discussed in Chapter 8.
- Representative plugging of core for analysis of different bed-types is especially important in thinly bedded reservoirs; see Chapter 7.
- Oil-based coring is recommended in cases where there is a need to obtain ground-truth core measurements of in-situ water saturation. Careful planning and execution is needed for such a coring operation to yield high-quality data.

**Note: We use a non-standard name and acronym (electrical borehole image, EBI) to refer generically to any one of a collection of micro-electrical borehole imaging tools. For reference to specific tools, see Chapter 9.*

- Longer, contiguous-cored intervals are strongly recommended (e.g., 90 ft [27 m] instead of 30 ft [9 m]). Core-to-log ties are often difficult in thinly bedded formations, and longer cored intervals facilitate making these ties.
- Thinly bedded intervals may have numerous partings or breaks due to fissile shales. Use core sleeves to provide better depth control, better recovery, and less chance of inverting sections of core when brought to surface.
- Obtain a core gamma-ray log at wellsite to aid in tying core to logs and to protect against inverting sections of core in subsequent handling. Alternatively, log the core gamma ray at the laboratory before slabbing and plugging.

Closely spaced sidewall cores

In lieu of conventional core, percussion or rotary sidewall cores can provide some insight into the thinly bedded nature of a prospective reservoir interval. In a hypothetical example of a 30-ft [9-m] interval with suspected thin beds (e.g., from a mud gas show), 30–60 evenly spaced, percussion sidewalls can provide a good statistical sampling of the reservoir interval. If very few or none of the sidewall cores contain sandstone, an alternative explanation for the gas show may be more probable; for example, a fractured shale.

Logging Programs for Thinly Bedded Formations

Conventional logs

The standard logging suite is not significantly changed for logging thinly bedded formations. Following are some general considerations for the selection and recording of conventional logs in thinly bedded formations.

- If logging conditions allow, use induction logs in preference to laterologs. All the modeling and analysis techniques discussed in this report are simpler and better validated for induction-type tools than for laterologs. This recommendation is especially important if there will be significant relative dip (greater than 30°) between the borehole and the bedding planes.
- If bed dip is significant, borehole-deviation and structural dip data are required to correct the induction log for dip effects.
- If possible, do not rely on acoustic logs for porosity. The effect of thin beds on acoustic tools is complex and variable and has not been quantified for log-analysis applications. High-quality density logs are preferred for determination of porosity.
- For each basic log measurement, use the highest-resolution log available from the selected service company.
- Record each log at the highest sample rate available — for example, 10 samples per ft instead of 2 samples per ft.
- Obtain whatever enhanced-resolution processing the service company offers for particular tools — for example, Schlumberger's 'alpha' processing for density and neutron logs [Flaum et al., 1989].

- Log each tool at a slow enough speed to support the high sample rate and enhanced-resolution processing options. Refer to the service company for specific recommendations.
- Consider running high-resolution conventional logs such as dielectric, dipmeter, or micro-log, especially if no image logs will be run. These tools can often define bed boundaries with more accuracy than gamma ray or porosity tools.

Specialty logs

- If possible, run an electrical borehole image log (EBI). See Chapter 9 for a list of these tools. In many cases these logs are invaluable for obtaining high-resolution bed definition. For additional information see Chapters 4 and 9.
- Consider running a Nuclear Magnetic Resonance (NMR) log. Experience shows that these logs may prove useful in confirming the presence of a thinly bedded reservoir [Chapter 4] and in estimating total porosity and irreducible water saturation in such formations. NMR measurements respond to thin beds with a simple volumetric-averaging characteristic which will allow them to be incorporated readily into the low-resolution analysis techniques described in Chapter 12.
- Consider running a multi-component induction log if it is available. As shown in Chapter 4, the component of resistivity perpendicular to bedding planes can be a good diagnostic indicator for the presence of a thinly bedded reservoir. Furthermore, in principle, assuming tool accuracy and modeling issues are resolved, the multi-component resistivity can significantly reduce the uncertainty in the determination of water saturation in the thinly bedded reservoir [Chapter 12].
- Be careful when creating long tool strings to reduce logging runs. Some tools can be combined for standard resolution, but data quality may be sacrificed if the tools are inappropriately combined. Check with the service company for specifics.

Summary

The acquisition of high-quality data for interpreting thin beds requires a good borehole. Water-based drilling muds should be formulated to avoid reactivity with formation shales. Often the best hole conditions can be obtained by drilling with oil-based or synthetic muds.

Core descriptions, core measurements and core photographs are essential elements for the rigorous evaluation of thin-bedded reservoirs. A suite of high-quality core photographs, taken in both white light and ultraviolet light, is a key tool. Representative plugging of core for analysis of different bed-types is especially important. The acquisition of core data and core photographs is discussed in Chapters 7 and 8.

The standard logging suite is not significantly changed for logging thinly bedded formations. However, it is recommended that data be acquired at the highest possible sample rate and that the highest-resolution available tools and data processing be utilized. Density and neutron logs should be used for porosity determination in preference to acoustic logs. Log modeling considerations suggest a preference for induction tools rather than focused-current contact tools, especially if there is significant relative dip between borehole and bedding planes.

An electrical borehole image tool should be run whenever thin beds are anticipated. A nuclear magnetic resonance (NMR) log can be very helpful both in confirming the presence of a thin-bedded reservoir and in quantifying hydrocarbon pore-thickness. Consideration should also be given to running a multi-component induction tool, a relatively new log which can potentially reduce the uncertainty in the estimate of sandstone water saturation in a thin-bedded reservoir.

**Lake Floor Fan,
Cretaceous,
Candeias Formation:
Itaparica Island, Bahia, Brazil**

Photo by Kevin Bohacs

6 Petrophysical Properties of Thinly Bedded Formations

Overview

Introduction

There are three main reasons for the difficulty of using well logs to evaluate hydrocarbon pore-thickness in thin-bedded reservoirs. First, well logs do not measure directly the reservoir properties of interest: The net sand thickness, sand porosity, and sand water saturation. Instead, the logs measure petrophysical properties of the rock such as bulk density and resistivity, from which the reservoir properties must be inferred. Second, the petrophysical log measurements represent *averages* over some collection of beds that are too thin to be measured individually. To infer reservoir properties from these measurements, therefore, it is necessary to understand how the properties of an extended reservoir volume are related to the properties of the individual bed types within that volume. The third reason is that some petrophysical properties are *anisotropic* — their values depend not only on the composition of the measured volume, but on a choice of orientation within that volume. Anisotropy further complicates the inference of reservoir properties from measured petrophysical properties. These issues are the topics of this chapter.

Contents

DOI: 10.1306/1157789A13220

Petrophysical Properties, Anisotropy, and Bed Types

Petrophysical properties and support

Petrophysical properties are intrinsic physical properties of a volume of rock, which have well-defined values independent of measurement techniques. Any reference to a measured value of a petrophysical property carries with it, explicitly or implicitly, a reference to the *support* of the measurement — that is, the physical volume of rock that is sensed by the measurement [Isaaks and Srivastava, 1989]. Typical petrophysical measurements have support volumes ranging from a cm^3 or less (e.g., measurements on drill cuttings or core plugs) to ten or more m^3 (e.g., resistivity measured by a deep induction log). The intrinsic vertical resolution of a borehole measurement is simply the vertical dimension of its support volume.

Isotropy and anisotropy

A petrophysical property that always depends only on its support volume, and on no other factor, is termed *isotropic*. Such a property always has a single value for a given rock volume and is independent of orientation. Total porosity, total water saturation, and bulk density are examples of isotropic petrophysical properties.

Some petrophysical properties, on the other hand, depend not only on the support volume but also on a direction or orientation of measurement. Such properties are termed *anisotropic*. Resistivity (or conductivity), permeability, and acoustic velocity are examples of anisotropic properties. For instance, we have already seen in Chapter 1 that the conductivity, or the resistivity, of a thinly bedded formation has different values when measured parallel or perpendicular to bedding planes. It is possible that a homogeneous, unlayered volume of rock may exhibit effectively isotropic values of conductivity, permeability, or velocity. However, even a shale-free, apparently homogeneous sandstone may exhibit conductivity anisotropy because of laminated variations in porosity [Kennedy and Herrick, 2004].

In this chapter we investigate how thin bedding induces anisotropy in the macroscopic conductivity and permeability of thinly bedded formations.

The complete mathematical description of an anisotropic petrophysical property requires the notion of a *tensor* quantity, which for a given coordinate system takes the form of a 3×3 matrix [Zhdanov et al., 2001]. In this publication we will mostly avoid the mathematics of tensors and will instead use simpler, more intuitive descriptions of anisotropic properties, always making explicit any simplifying assumptions involved in such descriptions.

Petrophysical bed types

In analyzing a thinly bedded formation, it is useful to classify the individual thin beds into two or more *petrophysical bed types*, which we can define as follows.

> A *petrophysical bed type* describes a group of petrophysical beds characterized by common, narrow distributions of porosity, grain density, permeability, and capillary-pressure behavior. A bed type is also expected to have narrow distributions of mineralogic composition and petrographic and petrologic attributes.

This definition will be discussed in greater detail in Chapter 7. Its significance in the present discussion is that, given such a classification, we can characterize the petrophysical properties of each bed type by their average values — we don't have to worry about the relatively minor variations among the different beds of a given type.

In the simplest case, we might divide a thinly bedded siliciclastic reservoir into sandstone and shale beds and characterize each of these bed types by its average properties.

Each of these average bed-type properties might be isotropic or anisotropic. For example, shale conductivity may be anisotropic. Often we make a working assumption that the properties of a given bed-type are isotropic; when such an assumption is used, it must be clearly stated.

Averaging Terminology and Definitions

Purpose

This section introduces the notation needed to express relationships between the petrophysical properties of individual bed types and the composite properties of a larger volume of a thinly bedded formation.

Definitions

All definitions refer to a fixed support volume, V, of the thin-bedded formation. Figure 6.1 illustrates such a volume with two bed types. This fixed volume might represent a section of core, a cylinder around the wellbore of given radius and height, or the investigation volume of a given logging tool.

- A *thin-bed property*, P, is the value of a petrophysical property with the volume V as support.
- A *bed-type property*, P_i, is the average value of a petrophysical property over that fraction of the fixed volume V composed of a given bed type, the i-th bed type.
- The *volume fraction*, v_i, of the i-th bed type is the ratio of the volume of that bed type to the total volume V (pore space included). If there are N bed types in V, the volume fractions satisfy

$$v_1 + v_2 + \cdots + v_N = 1.$$

- The *mass fraction*, w_i, of the i-th bed type is the ratio of the mass of that bed type to the total mass of the volume V (pore space included). If there are N bed types in V, the mass fractions satisfy

$$w_1 + w_2 + \cdots + w_N = 1.$$

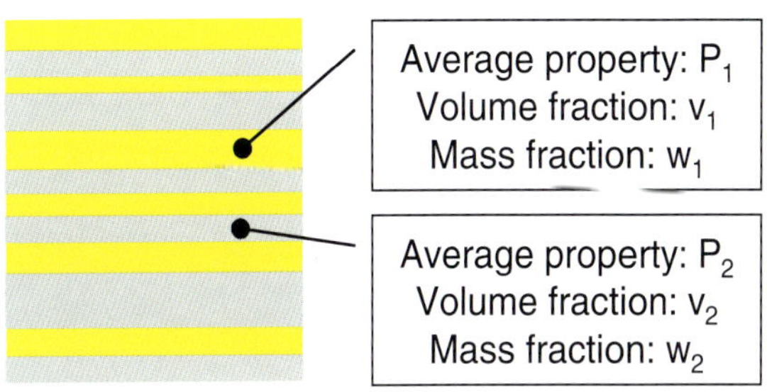

Figure 6.1. Thinly bedded formation with two bed types.

Isotropic Properties

Introduction

Many isotropic petrophysical properties are combined over a volume comprising several rock types by a simple volume-weighted average of the corresponding properties of the individual rock types. Such volume-weighted averages are independent of the geometry of the rock volume — they apply to thin planar beds, contorted beds, conglomerates, or any other arrangement. These volume-averaged properties include:

- The volume fraction of the i-th rock type, v_i
- Total porosity, ϕ_t
- Total water saturation, S_{wt}
- Bulk density, ρ_b, and matrix density, ρ_{ma}

Note: *It is important to identify the correct support volume for averaging each quantity. For example, as illustrated below, the total rock volume V is the support for averaging total porosity, while the total pore volume* $\phi_t \cdot V$ *is the support for averaging total water saturation.*

Total porosity

$$\phi_t = v_1 \cdot \phi_{t1} + v_2 \cdot \phi_{t2} \qquad \text{(2 bed types)} \qquad (6.1)$$

$$\phi_t = v_1 \cdot \phi_{t1} + v_2 \cdot \phi_{t2} + \ldots + v_N \cdot \phi_{tN} \qquad \text{(N bed types)}$$

Note: *For simplicity, the remaining equations will be presented only for the case of two bed types (like Equation 6.1).*

Total water saturation

Since total water saturation is defined as a fraction of total pore volume, it must be averaged over V using fractional pore volumes as weights:

$$\phi_t \cdot S_{wt} = v_1 \cdot \phi_{t1} \cdot S_{wt1} + v_2 \cdot \phi_{t2} \cdot S_{wt2} \qquad (6.2)$$

Bulk density

$$\rho_b = V_1 \cdot \rho_{b1} + v_2 \cdot \rho_{b2} \qquad (6.3)$$

Matrix density

Since matrix density is defined as a fraction of total matrix volume, it must be averaged over total rock volume V using fractional matrix volumes as weights:

$$(1 - \phi_t) \cdot \rho_{ma} = (1 - \phi_{t1}) \cdot v_1 \cdot \rho_{ma1} + (1 - \phi_{t2}) \cdot v_2 \cdot \rho_{ma2} \qquad (6.4)$$

Anisotropic Properties: Conductivity

Thin parallel planar beds

Rocks that have bedding on a scale finer than the support volume of electrical measurement will likely exhibit *macroscopic conductivity anisotropy*. Figure 6.2 illustrates a measurement volume comprising two thinly bedded rock types, with an (x, y, z) coordinate system superposed. The x and y axes in Figure 6.2 are parallel to the planar bedding and the z-axis is transverse to bedding.

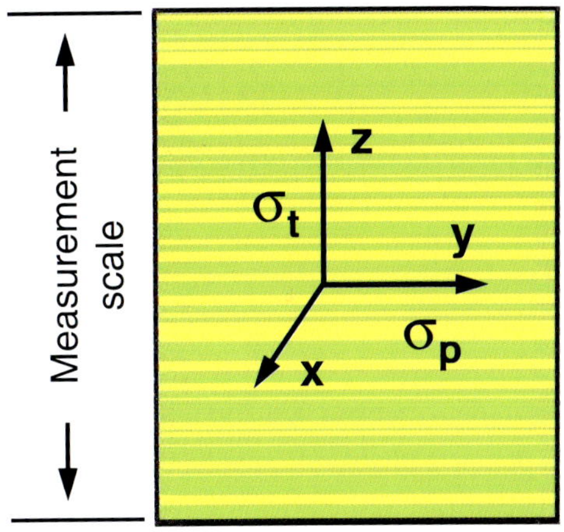

Figure 6.2. Anisotropic conductivity in thin-bedded formation.

The effective conductivity for current flow *parallel* to bedding planes (σ_p) differs from the conductivity for current flow *transverse* to bedding planes (σ_t) as shown in equations 6.5 and 6.6. Here the conductivities of the individual rock types (σ_1 and σ_2) are assumed isotropic.

$$\sigma_p = v_1\sigma_1 + v_2\sigma_2 \qquad \text{(parallel conductivity)} \qquad (6.5)$$

$$\frac{1}{\sigma_t} = \frac{v_1}{\sigma_1} + \frac{v_2}{\sigma_2} \qquad \text{(transverse conductivity)} \qquad (6.6)$$

These general averaging relationships, derived for ideal unidirectional current flow in planar beds of infinite lateral extent, are a simple consequence of Kirchoff's rules for combining resistive circuit elements in parallel and series [Poupon et al., 1954]. Thus, they do not depend on the theory of any particular resistivity logging tool. In Chapter 10 we will see that, when a borehole penetrates perpendicular to bedding planes, both induction and focused-contact resistivity logs essentially measure σ_p, the parallel conductivity; and we will also investigate what is measured when the borehole deviates from the perpendicular.

Figure 6.3 illustrates an example of parallel and transverse conductivities for thin parallel-planar beds, as given by Equations 6.5 and 6.6. The two rock types are sandstone and shale; shale conductivity is 1000 mS/m (corresponding to 1 ohm m resistivity) and the sandstone conductivity is 50 mS/m (corresponding to 20 ohm m resistivity).

Figure 6.3. Parallel and transverse conductivities for a thinly bedded formation.

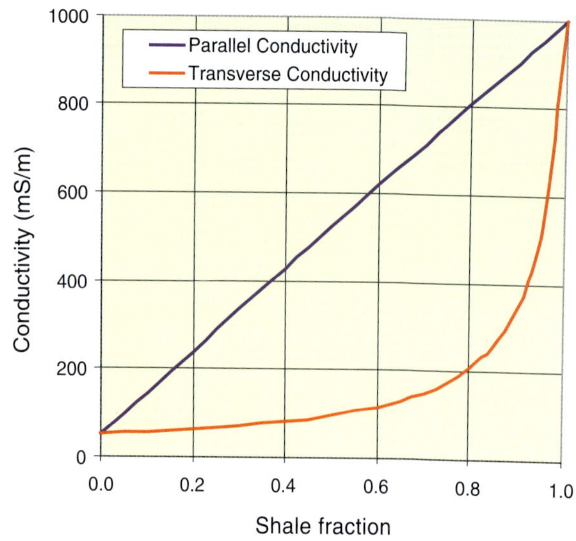

Non-planar beds

The anisotropic conductivity averaging equations, 6.5 and 6.6, are derived for a simple, highly idealized bedding geometry: Parallel, planar beds of infinite lateral extent. This is a fairly restrictive and somewhat unrealistic geometry, in contrast to the isotropic averaging equations (6.1 – 6.4), which are independent of the geometric arrangement of the constituent rock types. It is useful to investigate how the anisotropic average conductivity might differ under less restrictive geometric assumptions.

One way of describing a more general bedding geometry is to consider beds of finite lateral extent that are modeled as flattened ellipsoidal discs like those illustrated in Figure 6.4. These ellipsoids are characterized by a vertical semi-axis length, b_v, and a single horizontal semi-axis length, b_h. We can define the aspect ratio, α, of such an ellipsoid as the ratio of the vertical to the horizontal semi-axis:

$$\alpha = b_v / b_h \tag{6.7}$$

With this definition, planar beds can be considered the limiting case where $\alpha = 0$. Figure 6.4 illustrates the three cases $\alpha = 0.0$, 0.1, and 0.5 (not drawn to scale). Note that the individual "beds" are not all of the same size, but they all have the same aspect ratio and their principal axes are aligned in parallel.

While these ellipsoids still represent idealized bed-forms, we see that they can be used to approximate the geometry of thin interbedded sandstones and shales of finite extent (as in the figure with $\alpha = 0.1$) and shale clasts distributed within a sand body (as in the figure with $\alpha = 0.5$).

Figure 6.4. Thin beds of finite lateral extent.

It is possible to estimate the parallel and transverse conductivities for the class of geometries illustrated in Figure 6.4 by means of *effective medium theory* (EMT) [Dahlberg, 2002]. This method has been utilized by numerous authors to study the electrical and acoustic properties of sedimentary rocks [Mendelson and Cohen, 1982; Hornby et al., 1993].

The EMT approximation is not described here in detail. Instead, we present a graph (Figure 6.5) that shows the resulting parallel and transverse conductivities for sandstone and shale beds with the same internal conductivities as in Figure 6.3, and with aspect ratios ranging from 0.00 to 0.95. The legend in Figure 6.5 describes the curves in order from top to bottom (or left to right).

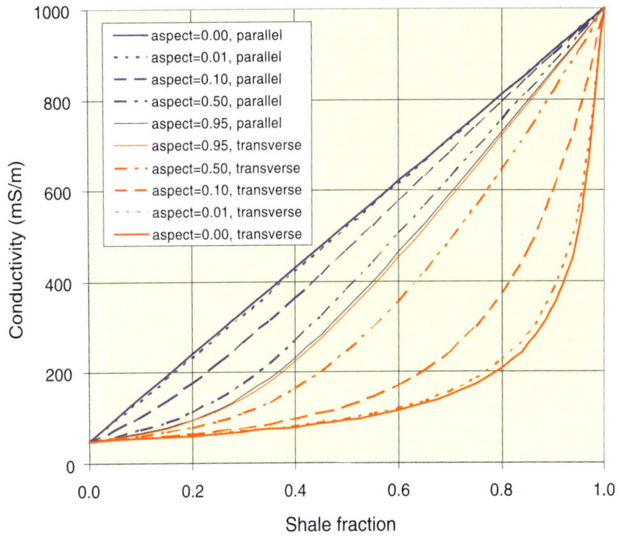

Figure 6.5. Parallel and transverse conductivities for thin beds with finite aspect ratios.

The curves with $\alpha = 0$ duplicate those in Figure 6.3 for planar beds. Note that as α approaches 1, the parallel and transverse conductivities converge toward a single curve which represents the isotropic effective conductivity for a sand body with spherical shale inclusions.

The method of *volumetric laminated sand analysis*, described in Chapter 12, incorporates this EMT approximation to estimate the effect of finite lateral bed dimensions in the determination of hydrocarbon pore-thickness.

Induction logs and the effect of dip

Induction logging tools measure conductivity via induced current loops concentric with the borehole. Therefore, the direction of current flow is perpendicular to the borehole, and the orientation of current loops relative to bedding planes depends on the relative dip angle, θ, between borehole and bedding planes as illustrated in Figure 6.6.

Figure 6.6 shows the apparent conductivity measured by an induction tool (σ_{IL}) and the parallel and transverse conductivities of the anisotropic formation (σ_p and σ_t). Moran and Gianzero [1979] determined the theoretical response of the induction tool to an anisotropic formation in the presence of dip. Equation 6.8 describes this ideal response in terms of θ, σ_p, and σ_t.

$$\sigma_{IL} = \sqrt{\sigma_p^2 \cos^2(\theta) + \sigma_p \sigma_t \sin^2(\theta)} \tag{6.8}$$

Figure 6.6. Induction tool in dipping thin-bedded formation.

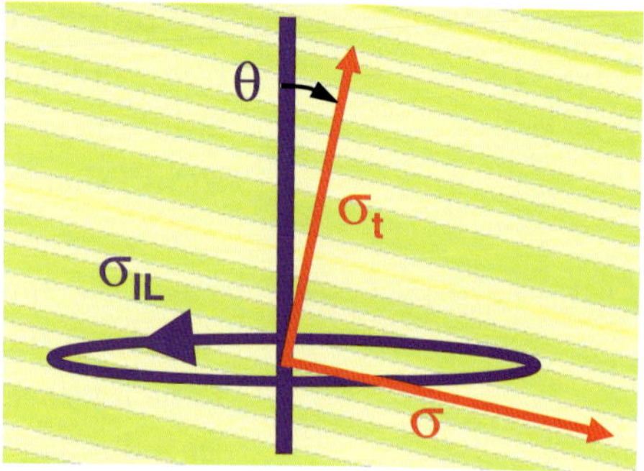

Equation 6.8 applies to any anisotropic formation, such as those illustrated in Figures 6.2 and 6.4, where there are only two independent values of conductivity — one (σ_p) that applies to all directions parallel to an orienting plane, and the other (σ_t) that applies to the direction orthogonal to this plane. Such a formation is sometimes called *transversely isotropic* [Kennedy et al., 2001]. The angle θ is the relative angle between the axis of the induction tool and the transverse, or σ_t, axis of the formation.

Note: Transverse isotropy and the conductivity tensor

It is always possible to construct a coordinate system that is aligned with the principal axes of a tensor quantity. In this coordinate frame the tensor is expressed as a diagonal matrix. In particular, the bedding geometries illustrated in Figures 6.2 and 6.4 are characterized by a conductivity tensor with only two independent components: The parallel or horizontal conductivity ($\sigma_p = \sigma_h$) parallel to the (x, y) plane, and the transverse or vertical conductivity ($\sigma_t = \sigma_v$) parallel to the z-axis. Thus in the (x, y, z) coordinate system the conductivity tensor is the simple diagonal matrix shown in Equation 6.9.

$$\hat{\sigma} = \begin{pmatrix} \sigma_h & 0 & 0 \\ 0 & \sigma_h & 0 \\ 0 & 0 & \sigma_v \end{pmatrix} \tag{6.9}$$

A formation whose conductivity tensor has only two independent components, as in Equation 6.9, is sometimes referred to as a transversely isotropic medium [Kennedy et al., 2001].

There is a simple relationship between the conductivity tensor in bedding-plane coordinates, Equation 6.9, and the corresponding tensor $\hat{\sigma}'$ referred to a different (x′, y′, z′) coordinate system. If the orthornormal rotation matrix R expresses the transformation from (x′,y′,z′) coordinates to (x, y, z) coordinates, as in the vector equation v = Rv′, then

$$\hat{\sigma}' = R^t \hat{\sigma} R \tag{6.10}$$

where R^t is the transpose of R. Equation 6.10 is useful, for example, when a borehole penetrates a thin-bedded formation at an angle deviated from the perpendicular to the bedding planes. The (x′, y′, z′) coordinate system can be aligned with an induction logging tool in such a borehole, and equation 6.10 can be used to relate the measurements made by such a tool to the formation conductivities σ_h and σ_v. This rotation, applied to the solution of Maxwell's equations for a magnetic dipole in a transversely isotropic medium, is the basis for Equation 6.8 [Moran and Gianzero, 1979].

Archie exponents

In a homogeneous rock, conductivity and water saturation are related by Archie's equation [Archie, 1942]. In equation 6.11, σ is the formation conductivity, σ_w is the brine conductivity, ϕ_t is total porosity, S_{wt} is total water saturation, and m and n are the cementation and saturation exponents (the Archie exponents):

$$\sigma = \sigma_w \cdot \phi_t^m \cdot S_{wt}^n \tag{6.11}$$

In a heterogeneous laminated sandstone formation, where porosity and grain size are not uniform, it is easy to assume that one may simply use average values of m and n in Equation 6.11 and obtain correct average water saturation for the composite formation. Such an assumption is incorrect, however; it does not account for conductivity anisotropy, and it may lead to significant errors in saturation.

Kennedy and Herrick [2004] present a detailed analysis of the effect of thin bedding on m and n. A key finding is that different values of m and n are required in Archie's equation depending on whether σ is the horizontal or the vertical conductivity, or something in between. A close look at Equation 6.11 makes this obvious: The scalar parameters σ_w, ϕ, S_{wt} are each isotropic, so for Archie's equation to be consistent with different values of formation conductivity in different orientations, it is necessary to change m and n with orientation.

In the methods developed in this publication, we do not apply Archie's equation (or its variants, the shaly sand saturation equations) in a direct calculation of the macroscopic water saturation of a composite thin-bedded formation. Instead we apply the saturation equations only to individual bed-types. We assume these bed-types are each homogeneous and isotropic to a sufficiently close approximation that we can apply single, non-directional values of m and n without introducing significant errors.

Anisotropic Properties: Permeability

Parallel planar beds

Consider a thin-bedded reservoir comprising permeable sandstone beds and impermeable shales. Clearly, the shales block fluid flow perpendicular to bedding, while flow parallel to bedding is not blocked. Thus permeability is an anisotropic rock property. The relationships for the horizontal (parallel) and the vertical (transverse) permeability for two interbedded bed-types are given in Equations 6.12 and 6.13 [Craft and Hawkins, 1959]. In these equations it is assumed that k_1 and k_2, the permeabilities of bed-types 1 and 2, are isotropic.

$$k_h = v_1 k_1 + v_2 k_2 \tag{6.12}$$

$$k_v = \frac{1}{v_1/k_1 + v_2/k_2} \tag{6.13}$$

Equations 6.12 and 6.13 are identical in form to the corresponding equations 6.5 and 6.6 for conductivity.

Correlations with resistivity

In a homogeneous hydrocarbon-bearing formation, high permeability tends to correlate with lower irreducible water saturation (S_{wirr}) and, hence, with higher resistivity. Conversely, lower permeability tends to correlate with higher S_{wirr} and, hence, with lower resistivity. Petrophysicists often make use of these correlations to estimate permeability from resistivity logs [Ahmed et al., 1989].

In a thin-bedded reservoir, such correlations are problematic because of the anisotropy of both resistivity and permeability. Qualitatively, the horizontal resistivity is dominated by the low-resistivity bed type (the shale) and the vertical resistivity is dominated by the high-resistivity bed type (the hydrocarbon-bearing sandstone). Conversely, the horizontal permeability is dominated by the sandstone while the vertical permeability is dominated by the shale. Thus, some authors have made correlations between horizontal permeability and vertical resistivity, and between vertical permeability and horizontal resistivity [Klein et al., 1997]. Correlations of this type must be evaluated carefully for each unique dataset; we do not use them in the methods developed in this text.

Summary

The *support* of a petrophysical measurement is the physical volume of rock that is sensed by the measurement and described by the measured value. A petrophysical property that depends only on the support volume of its measurement, and on no other factor, is termed *isotropic*. Porosity, water saturation, and bulk density are examples of isotropic properties.

A petrophysical property that depends not only on the support volume of the measurement but also on its orientation within the volume is termed *anisotropic*. Conductivity, permeability, and acoustic velocity are examples of anisotropic properties.

To understand how the average petrophysical properties of a heterogeneous volume of rock are related to the properties of the volume's constituents, it is useful first to subdivide the volume into a set of *petrophysical bed types*. A *petrophysical bed type* describes a group of petrophysical beds characterized by common, narrow distributions of porosity, grain density, permeability, capillary-pressure behavior, etc. The petrophysical properties of each individual bed type may be characterized simply by their average values.

An isotropic petrophysical property of a heterogeneous rock volume is related to the corresponding property values of its constituent bed types by a simple volumetric average. The appropriate volume fractions used as weights in the average depend on the support volume of the particular property. For example, the support for porosity is the total rock volume, the support for water saturation is the pore volume, and the support for grain density is the solid volume.

The anisotropic conductivity of a thin-bedded rock volume depends on the orientation of the measurement with respect to the bedding planes. If the current flow and the potential measurement are both oriented parallel to bedding, the resulting conductivity is a volume-weighted *parallel* average, analogous to resistors in parallel. This is the case for an induction logging tool in a borehole oriented perpendicular to bedding. On the other hand, if the current flow and potential measurement are perpendicular to bedding, the measured conductivity is a volume-weighted *series* average, analogous to resistors in series.

Similar averaging relationships exist for permeability, depending on whether the measurements of pressure difference and flow velocity are oriented parallel or perpendicular to bedding.

Deep-water,
Permian,
Brushy Canyon Formation:
Delaware Mountains, Texas, U.S.A.

Photo by Quinn Passey

7 Characterizing Thinly Bedded Reservoirs with Core Data

Overview

Introduction

This chapter describes the application of conventional core data in characterizing thin-bedded reservoirs. Cores and the analysis of core samples serve many objectives in the description and modeling of a hydrocarbon reservoir, from understanding the geology to establishing a quantitative model of reservoir flow characteristics. Here we focus on objectives related to the petrophysical evaluation of hydrocarbon pore-thickness (HPT) in thinly bedded reservoirs, where the primary data for evaluation are well logs and the core data are used to calibrate and/or validate the well log analysis.

Thin beds, and particularly very thin beds, present special challenges in the core-based characterization of the thinly bedded reservoir. Digital core images (i.e., digital versions of core photographs) have proven to be an invaluable tool in meeting some of these challenges. Applications of digital core images are mentioned several times in this chapter. The acquisition and analysis of these images is discussed in some detail in Chapter 8.

Contents

DOI: 10.1306/1157790A13220

Reservoir Characterization Objectives

Overview

Cores and the analysis of core samples serve many functions in the description and modeling of a hydrocarbon reservoir, from understanding the geology to establishing a quantitative model of reservoir flow characteristics. Here we focus on objectives related to the petrophysical evaluation of hydrocarbon pore-thickness (HPT) in thinly bedded reservoirs, where the primary data for evaluation are well logs and the core data are used to calibrate and/or validate the well log analysis. We also emphasize the importance of establishing a petrophysical model for HPT analysis that is consistent with the geologic description of the reservoir.

Objectives

Specific objectives are listed and described in Table 7.1.

Table 7.1. Objectives of core analysis in a thinly bedded reservoir.

Objective	Description
Establish petrophysical model	The petrophysical characterization of a thin-bedded reservoir starts with a model comprising a complete set of *petrophysical bed types*. To be complete, the petrophysical model must cover the whole range of reservoir quality: Reservoir, marginal reservoir, and non-reservoir rocks. Thus, the first objective for core analysis is to collect a sufficiently representative sample dataset to enable the identification and characterization of all the petrophysical bed types.
Characterize bed types	Log measurements in a thin-bedded reservoir inevitably represent averages of the properties of all the bed types present. The function of log analysis, then, is to separate out the properties of the reservoir beds from this mix. In order to utilize core data as ground truth for log analysis models, it is necessary first to characterize the average petrophysical properties of each of the individual *petrophysical bed types* at a scale comparable with log measurements. The properties required include total porosity and grain density, and for reservoir bed types, the primary-drainage capillary pressure curve and the electrical properties (e.g., Archie equation exponents).
Determine macroscopic reservoir properties	In order to compare core-derived data directly to well logs, it is necessary to combine the core data for bed types in macroscopic averages that represent the properties of the reservoir on the scale of well-log measurements. Particular care is required to combine these properties correctly based on the distribution of bed types, and digital analysis of core images is an important tool here [Chapter 8].

Core Preparation and Sampling Procedures

Overview

Obtaining high-quality core data to satisfy the petrophysical objectives described above depends on the successful completion of many complex steps in the drilling, handling, preservation, transportation, sampling, and measurement of the core. Various groups — e.g., geologists and reservoir engineers — have objectives for the coring operation that may differ from the petrophysicists' objectives. Coordinating these differing objectives requires careful planning of the coring and core analysis operations.

It is outside the scope of this publication to discuss these issues in detail. Here we give only a brief summary of an operational protocol that is employed by ExxonMobil to acquire, preserve, and measure cores from deep-water offshore wells [summary and photos courtesy of J. S. Campbell]. Many aspects of this particular protocol are unique to coring operations in unconsolidated sediments. For a more detailed discussion of issues related to quantitative measurements on cores, see Basan et al. [1988].

At the wellsite

Inner core barrels are either frozen or epoxy-stabilized on the rig, then cut into 1-m [3.3-ft] sections. The ends are capped and taped. Sections are packed horizontally to prevent gravity drainage of core fluids; see Figure 7.1. Special cushioned crates are used for shipping.

Core gamma ray

On arrival at the core analysis lab, core sections are inventoried and then processed through a core gamma machine (Figure 7.2). Core gamma-ray (GR) data are commonly collected on unslabbed cores, and allow the core data to be tied to log depth. Even in thinly bedded reservoirs, low-resolution GR data can be useful for correlating core and log. Most core gamma machines measure total gamma radiation. Some labs have spectral core gamma machines that measure potassium, thorium, and uranium.

Figure 7.1. 1-m core sections capped and packed for shipping.

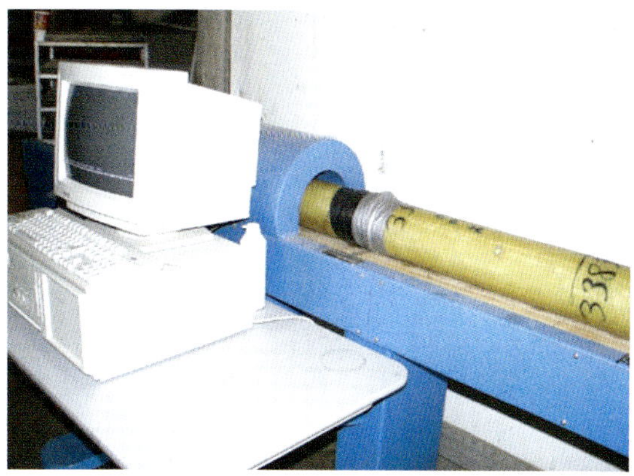

Figure 7.2. Core sections being processed through core gamma machine.

Whole-core CT scans

After obtaining the core GR, the core tubes are scanned for Computed Tomography (CT) before slabbing. Typically two longitudinal CT scans are taken at 90° orientations in addition to several cross-sectional CT scans per meter. The whole core CT scans are used to assess overall core quality, missing core intervals in the tubes, damaged or fractured intervals, and bed dip orientation. Figure 7.3 shows a core tube entering the CT scanner, and Figure 7.4 shows a set of cross-sectional scans.

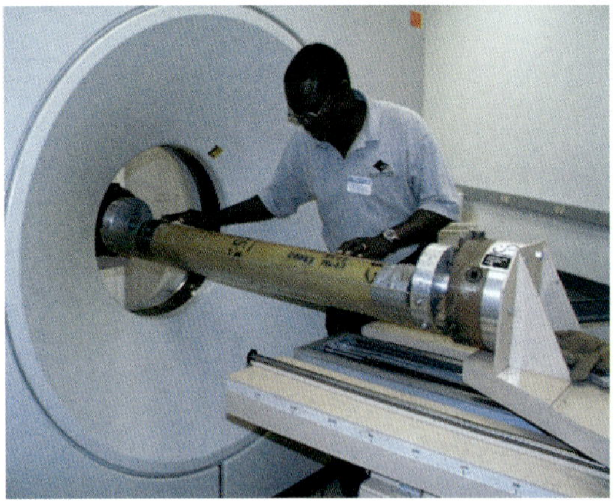

Figure 7.3. Core tube entering CT scanner.

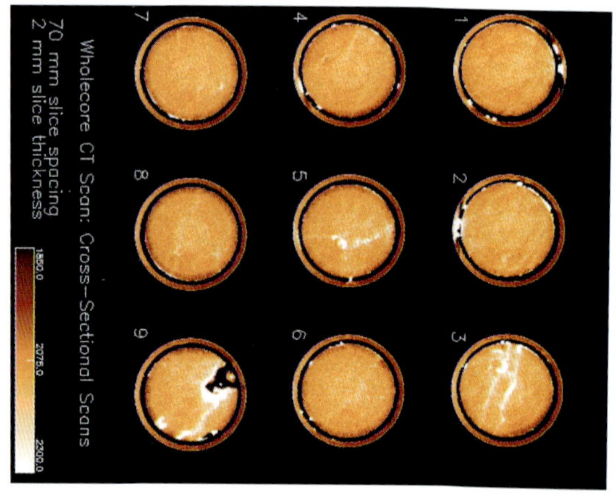

Figure 7.4. Cross-sectional CT scans from a single core tube.

Preservation and slabbing

After CT scanning, the core tubes are frozen before slabbing, if not previously frozen on the rig. Whole-core CT scans are reviewed to determine optimum slab orientation. The core tubes are always slabbed with maximum exposed bed dips to ensure the core plugs are obtained parallel to the bed strike and not perpendicular to bed dip, as illustrated in Figure 7.5.

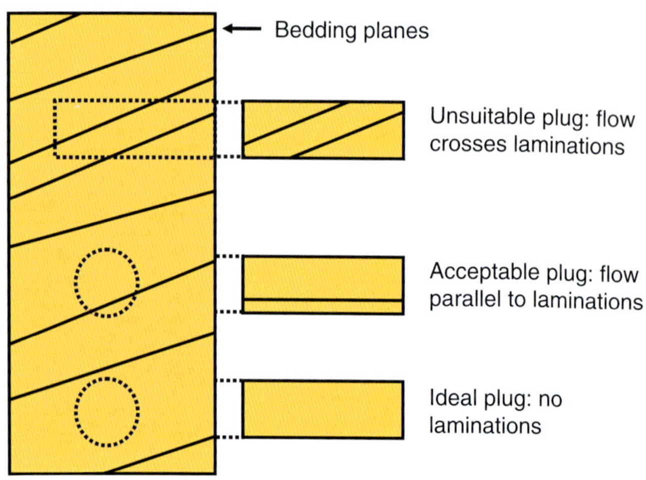

Figure 7.5. Core slabbed with maximum exposed dip to allow plugging perpendicular to slab face. Plugs should not cross bedding planes.

The frozen core tubes are slabbed with a radial band saw using liquid nitrogen as a coolant (Figure 7.6). The core tubes are oriented for slabbing based on the CT scans and are slabbed one-quarter–three-quarter (Figure 7.7). The one-quarter slab is used for detailed facies descriptions, white light and ultraviolet photography [Chapter 8], and probe permeametry measurements. The three-quarter slab is used for obtaining all core plugs for both routine and special core analyses.

Figure 7.6. Slabbing frozen core with radial band saw.

Figure 7.7. Core slabbed in one-quarter and three-quarter slabs.

Photography and geologic description

Immediately after slabbing the core tubes, the one-quarter core slabs are thawed and photographed (normally half-scale) both in white light and ultraviolet (UV) light. Cores are then grouped into 5-m [16-ft] intervals for white-light and UV "box photos." After the three-quarter core slabs are plugged, they are photographed in white light to document where all core plugs were obtained. Chapter 8 discusses the application of core photography in the evaluation of thinly bedded reservoirs.

After the one-quarter core slabs are photographed, the geology team performs a detailed facies description of the entire core (Figure 7.8). The petrophysicist works with the team integrating wireline log responses to help fully characterize the reservoir quality of the various facies.

Figure 7.8. Describing facies in the slabbed core.

For sufficiently well-consolidated cores, densely-sampled permeability measurements may be made on the one-quarter core slabs using the unsteady-state probe permeametry technique. When included, this step comes after the cores have been described.

Plugging and plug selection

Individual 1.5-in. [3.8-cm] diameter core plugs are cut from the frozen three-quarter slab using a diamond tipped core barrel and liquid nitrogen as coolant. The frozen core plugs are extracted from the three-quarter slab and are wrapped in plastic wrap, placed in individual containers, and then placed on dry ice to remain frozen. The practice in deep-water cores is to plug at high density, approximately 15 plugs per m in reservoir intervals and 6 plugs per m in shale sections (Figure 7.9) The high-density plugging ensures that an adequate number of plugs will be available for both routine and special core analyses. In some deep-water cores, up to 30–40% of the plugs are damaged and cannot be used for measurements.

Figure 7.9. High-density plugging employed in deep-water cores.

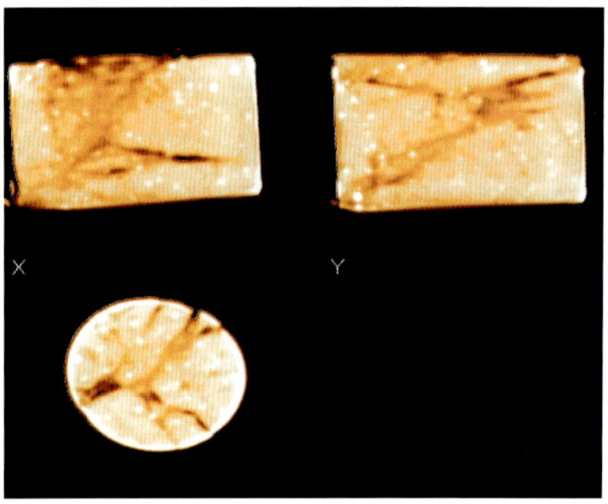

Figure 7.10. Longitudinal and cross-sectional CT scans of a core plug. Note the fractures.

In reservoir intervals, a 4–6 in. [10–15 cm] section of the three-quarter core slab is retained every other m to obtain a vertical plug and companion horizontal plug. The vertical plugs are used for rock mechanics and compressibility analyses that are critical in reservoir flow simulations and also to obtain the vertical–horizontal permeability (K_h–K_v) relationships. The vertical plugs are not used in any of the routine core analyses studies.

After the plugs are cut, they are scanned for computed tomography (CT) while frozen. Similar to the whole-core CT scans, the core plugs are CT scanned in two longitudinal orientations at 90° and one cross-sectional CT scan per plug (Figure 7.10). The plug CT scans are critical in selecting which plugs are used in the routine and special core analysis tests. Fractured and damaged plugs are eliminated from any testing.

An integrated team comprising geologists, petrophysicists, core analysts, and reservoir engineers meets to select plugs for both the routine and special core analysis programs. The team uses the whole-core CT scans, plug CT scans, core gamma, detailed core descriptions, profile permeametry, and the depth-tied wireline log responses to select representative plugs that are in good condition and suitable for each type of test. It is important for all the parties who will be involved with the core analysis testing and who will be using the results from the core analyses to be involved in the selection of the plugs and core-analysis program.

The measurement strategy for the characterization of a thin-bedded reservoir is outlined below. Descriptions of the various core testing procedures can be found in Basan et al. [1988].

Measurement Strategy

Representative sampling

The goal of core-plug sampling should be to obtain a statistically representative subset of samples to characterize the distribution of properties for each *petrophysical bed type*. In general it may be necessary to sample the core before the bed-type classification is established, since the core dataset itself is a primary tool for establishing such a classification. Thus a scheme of representative sampling, like that illustrated in Figure 7.11, should be used. In contrast, *uniform sampling* (where plugs are taken at roughly constant spacing) is not adequate. The practice of *high-graded sampling*, where plugs are selectively cut from the best-looking reservoir intervals, should be assiduously avoided.

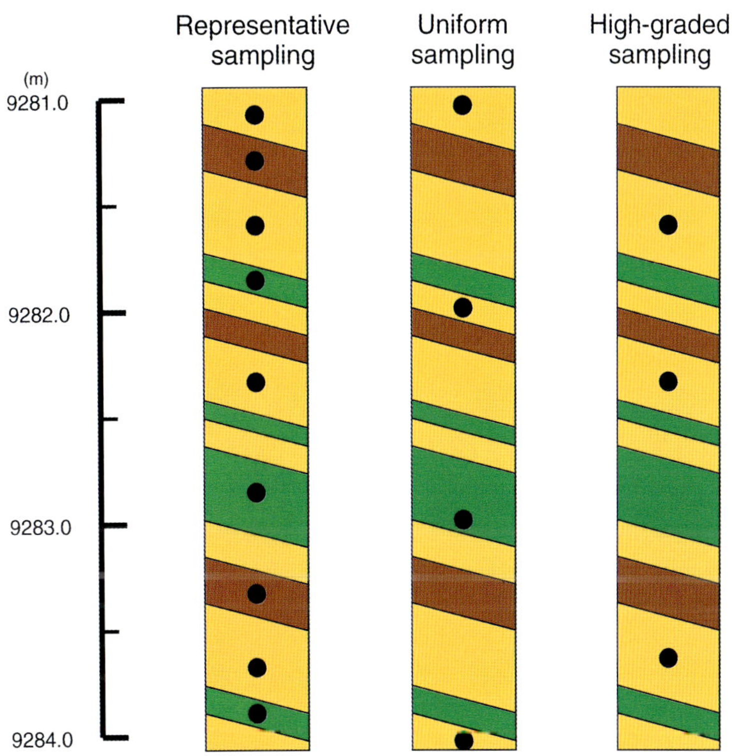

Figure 7.11. Core sampling strategies. For thin beds, representative sampling is recommended.

Note that the core plug data are to be used to characterize the petrophysical properties of each bed type, but are not to be used to estimate the distribution of bed types — that is, the proportions of each bed type present within the cored interval. The distribution of bed types is best determined either from the detailed core description or by the analysis of digital core photographs [Chapter 8].

Following are a few additional sampling guidelines.

- Core plugs should be homogeneous (avoiding bed boundaries).
- Plugs of 1-in. [2.5-cm] diameter are adequate for conventional log analysis (e.g., porosity, permeability, saturations, electrical properties).
- Plugs of 1.5-in. [3.8-cm] diameter should be collected for special core analyses (e.g., relative permeabilities, wettability, etc.).

Vertical plugs can be obtained in the thicker beds for vertical permeability determination, as needed.

Core plug measurements

Listed in Table 7.2 are the measurements needed to fully characterize each sample, in decreasing order of their importance for the classification and characterization of petrophysical bed types. This order may vary in some applications.

Table 7.2. Core measurements to characterize petrophysical bed types.

Measurement	Comment
Routine core analysis (1)	Reservoir and non-reservoir beds
Primary-drainage capillary pressure curve (2)	Reservoir beds
NMR T_2 distribution	Reservoir beds, if NMR logging planned
Electrical m, n, C_o/C_w	Reservoir beds (C_o/C_w as needed)
Native-state resistivity	Non-reservoir beds
Quantitative mineralogy (X-ray diffraction and X-ray fluorescence)	Reservoir and non-reservoir beds
Petrography (point-count)	Reservoir and non-reservoir beds
Grain size	Reservoir beds

(1) Porosity, permeability, grain density
(2) Preferred: Centrifuge; next best: Porous plate; next best: Mercury injection

Digital core images

Core photography is a vital part of the core measurement strategy for characterizing a thinly bedded reservoir, and digital capture of the photographs is essential. Here are several applications of digital core images:

- To determine net reservoir fraction (net-to-gross ratio);
- As the basis for identifying petrophysical bed types at very high resolution, determining fractions of each bed type over macroscopic intervals, and calculating average reservoir properties over such intervals;

- To generate bed-thickness statistics;
- To generate a high-resolution log-to-core depth match by aligning photographs with borehole image data;
- To develop ground-truth reservoir data for comparison with log-analysis results.

The methods and applications of digital core imaging are discussed in Chapter 8.

Digital capture of core description

The detailed geologic description of cores is an invaluable tool for integrating the petrophysical analysis with the geologic model of the reservoir. For this integration it is vital to capture the core description in digital format. Figure 7.12 illustrates a manually documented core description (on the left) that has been captured in digital format and redisplayed (on the right). Colors in the digital display represent different lithofacies. The horizontal scale is grain size.

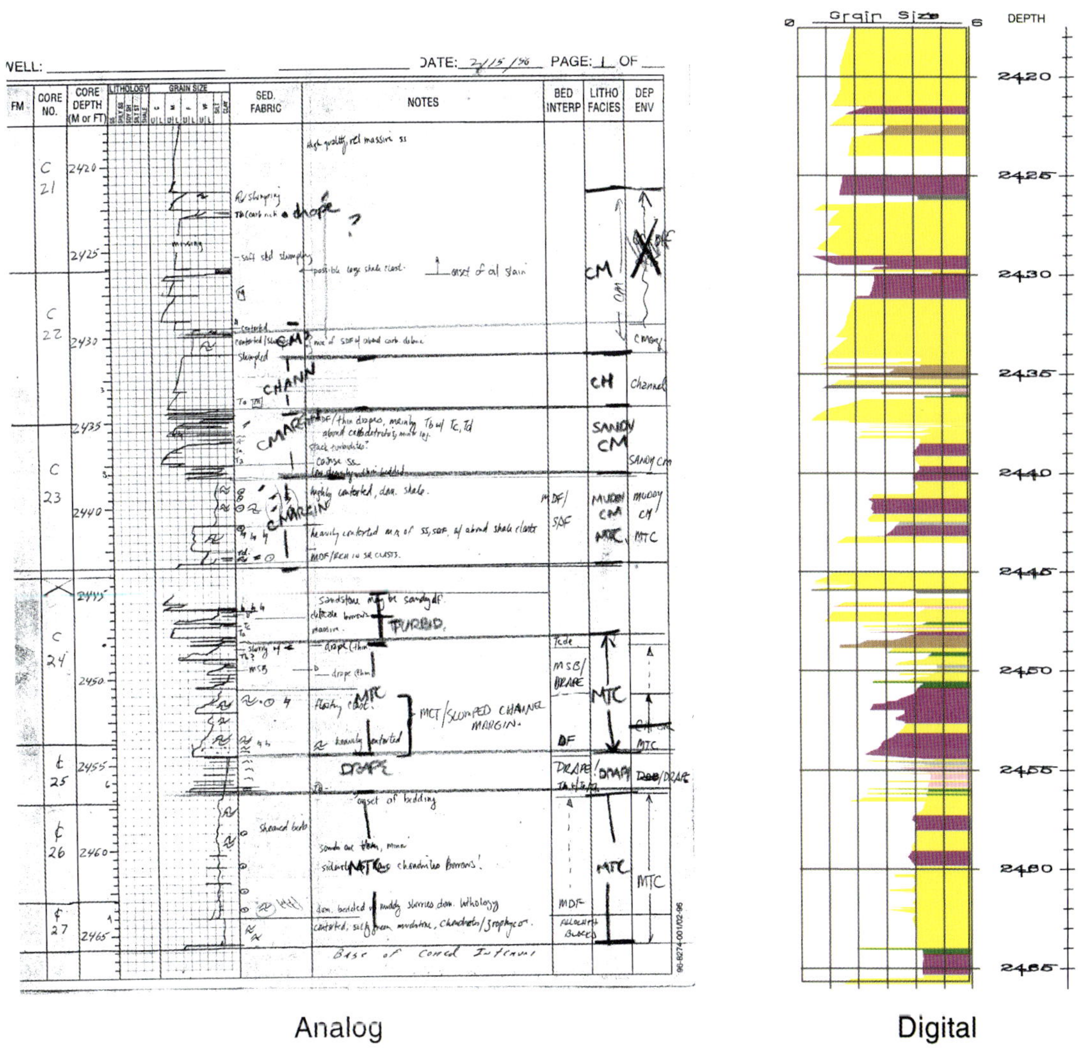

Figure 7.12. Analog core description (left) and digital version (right).

The digitized core description, integrated with digital core plug data and well log data, can be used to:

- Define *petrofacies* and calibrate well-log models for identifying petrofacies in uncored intervals [Sullivan et al., 2003];
- Define *petrophysical bed types* consistent with petrofacies;
- Quantify the bed-type distribution over the cored interval;
- Align core depths with log depths.

Characterizing Very Thin Beds

The problem

As defined in Chapter 3, petrophysical *thin beds* range in thickness from 1 in. [2.5 cm] to 2 ft [.6 m]. Conventional core plugs can be cut from these beds. On the other hand, petrophysical *very thin beds* are less than 1 in. [2.5 cm] in thickness and cannot be plugged.

When a reservoir has a mix of thin and very thin beds, it is important to question whether core plug data from thicker beds are adequate to characterize the average properties of the reservoir bed types. There may be a dependence of reservoir quality on bed thickness, such that (for example) very thin beds may be lower in porosity or higher in irreducible water saturation compared to thicker beds. A geologic understanding of the depositional environment of the reservoir may shed light on this issue.

Figure 7.13 shows an 86-ft [26-m] section of core photographs from a gas reservoir with mixed thin and very thin beds. Clearly it would be risky to assume that the average properties of this reservoir may be adequately characterized by core plug measurements from the thicker beds alone.

Figure 7.13. An 86-ft [26-m] section of Vicksburg formation in which many beds are below core-plug resolution. The inset (left) shows one bed where homogeneous core plugs could be obtained.

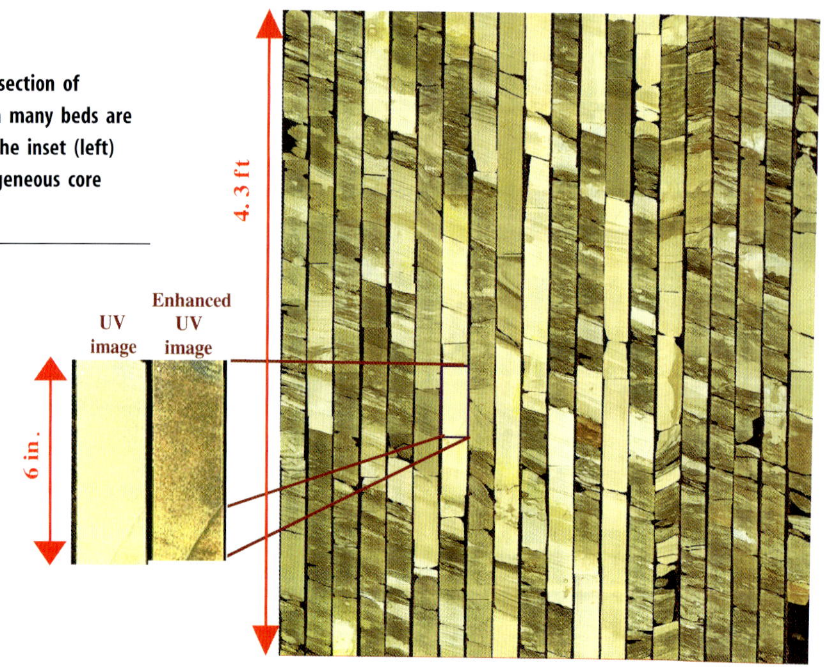

Digital core images

Digital core images such as Figure 7.13 are especially valuable when very thin beds are present. See Chapter 8 for details on the acquisition and analysis of digital core images.

1) The image-based determination of net reservoir thickness, or more generally the quantitative distribution of different bed types, is not limited by the presence of very thin beds.

2) Digital image analysis can be applied to measure the thickness of each bed throughout the imaged interval and derive a bed-thickness distribution for each bed type. Knowledge of this distribution allows us to evaluate the potential impact of the very thin beds on the average properties of a bed type. For example, if only 10% of the beds are very thin, their impact on the average porosity is much less than if they comprise 50% or more of the total count for a bed type.

Quantitative measurements on small samples

It is possible to measure porosity and grain density on small samples cut from the core slab. Quantitative mineralogy based on X-ray diffraction and X-ray fluorescence does not require plug-sized samples. Dahlberg and Fitz [1988] describe the use of these small-sample techniques in a thinly bedded reservoir to characterize log-scale average properties. More generally, the same techniques can be applied to small slab samples cut from individual bed types. Mercury injection capillary pressure measurements may also be made on small well-consolidated samples cut from the slab.

It is important to test the quality of these small-sample measurements. For this purpose, small samples can be cut from the ends of core plugs taken in thicker beds, and the small-sample measurements can be compared to the standard measurements on the corresponding plugs.

Profile permeametry

Profile permeametry is a commercially available measurement using a device illustrated schematically in Figure 7.14.

The profile permeameter operates as follows:

1) A 2-mm to 4-mm [.07 to .15-in.] diameter probe head is applied to the slabbed surface,

2) A 10-psi nitrogen pulse is applied,

3) The pressure decay as a function of time is monitored, and

4) Permeability is calculated from steady-state hemispherical flow.

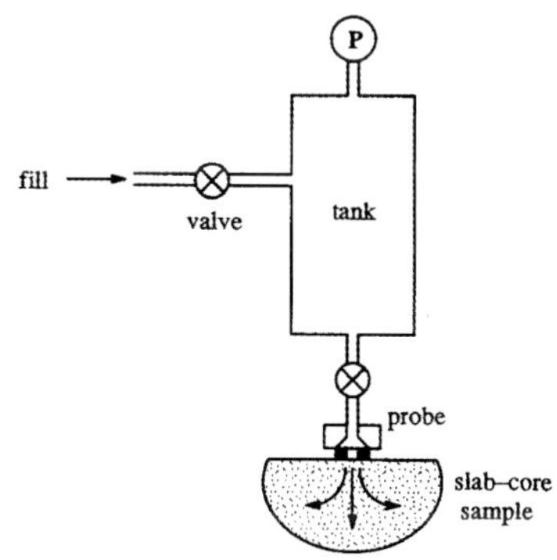

Figure 7.14. Schematic of profile permeameter applied to core slab.

Profile permeameter measurements are generally fast (30–40 measurements/hr), accurate, reproducible, and non-destructive. Their main drawback is that data quality tends to be poor in unconsolidated or low-permeability rocks.

Figure 7.15 shows a 0.9-m [3 ft] core interval comprising very thin beds from a tidal depositional environment. Because many of the sand beds in this formation are thinner than 1-in. [2.5-cm] diameter core plugs, a profile permeameter was used to help characterize the interval.

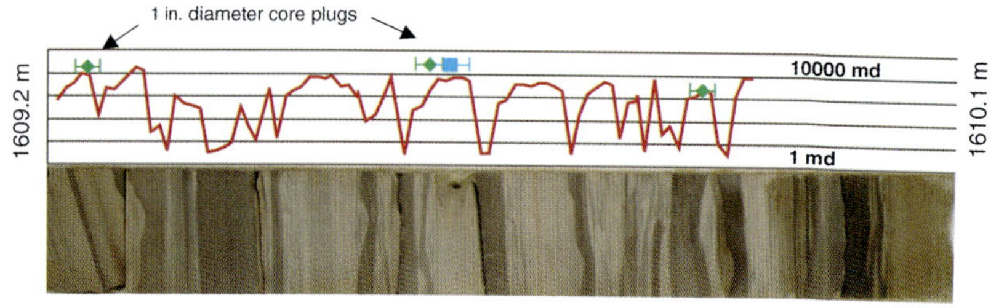

Figure 7.15. Photograph and high-resolution permeability profile of laminated interval.

The red curve shows the permeability profile obtained from the permeameter (using a 2-mm diameter probe). The permeability profile is in good agreement with core-plug permeability values obtained in the thicker beds (green and blue points). The bars on the plug values indicate the diameter of the plugs. The vertical sampling rate for the permeability profile was set at 0.8 cm (i.e., yielding 3.2 measurements/in.) thus allowing for characterization of beds much thinner than the diameter of routine core plugs.

Petrophysical Bed Types

Definition

In modeling a thin-bedded reservoir, it is a useful simplification to categorize the beds into a small number of discrete *petrophysical bed types*. Such a classification is required by some analysis methods (Chapter 12) and is a helpful constraint for others (Chapter 11). Thus, for these purposes, a primary objective in sampling and analyzing core is to identify and characterize these *petrophysical bed types*, which we define (somewhat loosely) as follows.

> A *petrophysical bed type* describes a group of petrophysical beds characterized by common, narrow distributions of porosity, grain density, permeability, and capillary pressure behavior. A bed type is also expected to have narrow distributions of mineralogic composition and petrographic/petrologic attributes.

In a given geologic environment it is unlikely that a unique way to classify the constituent rocks into petrophysical types would be found. However, it is important to establish a classification consistent with the geologic description of the reservoir. This consistency facilitates the integration of the petrophysical analysis into the geologic and resevoir-engineering models of the reservoir.

Relation to petrofacies

One way to ensure the consistency of the petrophysical bed types with the geologic description of the reservoir is to subordinate the definition of bed types to a *petrofacies* classification. As defined by Sullivan et al. [2003],

> *Petrofacies* are high-order lithofacies associations from detailed core descriptions that are grouped by similarity of log response.

Petrofacies, then, are themselves subordinated to the geologic description, and are defined at a scale characterized by well-log measurements. This scale makes it possible to attempt to predict petrofacies in uncored wells by well-log classification methods such as discriminant function analysis [Sullivan et al., op. cit.]. However, the scale also implies that a given petrofacies may comprise more than one *petrophysical bed type*. For example, the various turbidite petrofacies discussed by Sullivan et al. [op. cit.] are thin-bedded reservoir facies and would need to be subdivided, at a minimum, into sandstone and shale petrophysical bed types.

Classifying bed types

When a representative core database has been obtained, one can identify a set of petrophysical bed types by clustering these core data into a set of narrow distributions. Following are some general, if sketchy, guidelines for this process.

1) The number of bed types should be as small as possible.
2) Different bed types must have different responses on one or more well logs; otherwise they cannot be distinguished by log analysis.
3) A bed type must not include separate beds that have widely varying log responses.

Examples for each guideline follow.

1) In the simplest case, a thinly bedded formation may consist of interlayered sandstone and shale beds, where the sandstone beds all have uniform petrophysical characteristics and the shale beds likewise have uniform characteristics. In this case, there are only two petrophysical bed types.
2) If the sandstone beds vary in grain composition, e.g., in feldspar content, but otherwise have similar porosity, permeability, and capillary pressure behavior, then they probably have very similar responses on porosity and resistivity logs. In this case, they probably should not be subdivided into two classes based on grain composition alone.
3) If the spread between the minimum and maximum irreducible water saturation of the sandstone beds exceeds 20–30 saturation units, then these beds should probably be subdivided into two types based on a saturation cutoff. While this specific guideline is arbitrary, some such subdivision may be necessary for the low-resolution log analysis techniques described in Chapter 12. Similarly, shales and tightly cemented sandstones usually would not be lumped into a single non-reservoir bed type, because they would likely have different density and conductivity responses.

Characterizing bed types

When a bed-type classification has been established, the next step is to characterize the statistical distributions of the petrophysical properties of each bed type. For most purposes these distributions are adequately characterized by the mean, standard deviation, minimum, and maximum of the sample values. Ideally, these statistics for porosity, irreducible water saturation, grain density, and electrical properties will be obtained from a representative set of core-plug-based measurements for each bed type. When a significant proportion of *very thin beds* are present it may be necessary to include data from measurements made on smaller samples.

For purposes of uncertainty analysis, it is not enough to derive just the simple statistics of the sample dataset as mentioned above. It is necessary also to estimate the uncertainty in the mean value of each petrophysical parameter. This subject is discussed in the *Note*, page 86.

For a thorough and rigorous formation evaluation of a thinly bedded formation, it is important to gather core data from non-reservoir as well as reservoir bed types. In particular, native-state resistivity measurements on shale plugs can be useful to characterize the electrical properties of the shale beds and, thereby, to help analyze water saturations in the reservoir beds. Sometimes, however, the petrophysicist arrives on the scene to find that non-reservoir bed types have been neglected in core plugging and analysis. In such cases it is necessary to obtain (for example) shale porosity and conductivity from log readings in thick shales, and to assume that these properties are the same in the thinly bedded shales.

Note: Uncertainty in bed-type mean values

To characterize the uncertainty in bed-type parameters such as porosity and irreducible water saturation, we need to understand the uncertainty in the mean value. *For example, if the mean sand porosity is estimated as the average of a representative database of core-plug porosity values, then the uncertainty in this mean porosity is estimated by the* standard deviation of the mean. *This statistic is different from, and smaller than, the sample standard deviation of the set of plug porosity values. Refer to standard statistical texts, such as Draper and Smith [1982], for a discussion of the standard deviation of the mean. The following example illustrates the estimation of the standard deviation, minimum, and maximum of the mean sand porosity from a representative set of core plug measurements.*

The hypothetical core-plug porosity data set summarized below represents a high-quality sandstone bed type.

GIVEN:

N = *sample size* = *31 plugs*

ϕ_{min} = *minimum plug porosity* = *.24*

ϕ_{max} = *maximum plug porosity* = *.35*

ϕ_{mean} = *sample mean porosity* = *.29*

σ_{sample} = *sample standard deviation* = *.03*

THEN:

σ_{mean} = standard deviation of the mean = $\sigma_{sample}/\sqrt{N}$

$= .03/\sqrt{31} \cong .005$

Given these data, we can estimate the minimum and maximum possible values of the mean sand porosity. The range of values for the mean sand porosity is considerably smaller than the range of the individual plug porosity values (0.24 to 0.35). The approximate ninety-five percent (95%) confidence bounds for the mean are:

$$MIN_{95} = \phi_{mean} - 2 \cdot \sigma_{mean} = .29 - .01 = .28$$

$$MAX_{95} = \phi_{mean} + 2 \cdot \sigma_{mean} = .29 + .01 = .30$$

Macroscopic Reservoir Properties

Bed-type fractions

To determine the averages of the petrophysical properties over a macroscopic cored interval, the first step is to determine the fractions of each of the petrophysical bed types present. This may be accomplished in one of two ways: First, by a detailed manual description and bed-by-bed measurement of the core itself; or second, by a similar description and measurement carried out on digital core photographs. The classification and measurement of beds in core photographs may be at least partially automated by applying digital image analysis techniques. See Chapter 8 for examples. Figure 7.16 illustrates the process of estimating bed-type fractions by measuring the combined thicknesses of beds of each type and dividing by the total length of the cored interval.

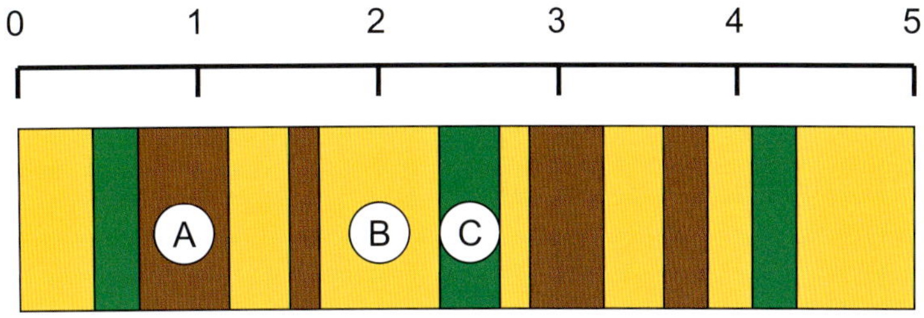

Figure 7.16. Determining bed-type fractions from thickness measurements.

Bed type:	A	B	C	Interval
Thickness	1.34	2.82	0.84	5.00
Fraction	0.27	0.56	0.17	1.00
Porosity	0.12	0.29	0.18	**0.225**

Macroscopic reservoir properties

Chapter 6 presents equations for combining bed-type petrophysical properties to obtain averages over macroscopic reservoir intervals. For example, Figure 7.16 illustrates the averaging of total porosity. The interval-average porosity in the lower right cell of of the table is calculated as

$$\phi = v_A\phi_A + v_B\phi_B + v_C\phi_C$$

where the terms v_x and ϕ_x are the volume fractions and porosities of the individual bed types.

Summary

For the petrophysical evaluation of HPT in a thin-bedded reservoir, conventional core analysis data serve three broad objectives. First, the core data are used to establish a petrophysical model comprising a complete set of *petrophysical bed types*. Second, the core data are used to characterize the average petrophysical properties of each bed type. Third, the core data (possibly combined with digital core image data) are used to determine the macroscopic properties of continuous cored intervals (e.g., net reservoir thickness, average porosity, and average water saturation) for direct comparison with the results of log analysis.

Geoscientists and reservoir engineers depend on core data for numerous other reservoir-description objectives. Obtaining high-quality data to satisfy all these objectives depends on the successful completion of many complex steps in the drilling, handling, preservation, transportation, sampling, and measurement of the core. An appropriate operational protocol is essential.

The goal of core-plug sampling is to obtain a statistically representative subset of samples to enable the identification of bed types and to characterize the distribution of petrophysical properties within each bed type. Detailed core descriptions or digital core photographs may be used to determine the volumetric distribution of bed types within contiguous cored intervals. For this application, the core descriptions should be captured in digital form.

The characterization of *very thin beds* (less than 1 in. [2.5 cm] thick) poses special challenges since these beds cannot be plugged homogeneously. It is risky to assume that very thin beds have the same properties as thicker beds; for example, there might be an inverse relationship between bed thickness and reservoir quality. Digital core images can be used to determine the distribution and importance of very thin beds. Some quantitative measurements can be made on small samples cut from the slabbed core. Profile permeametry, applied to the slabbed core, can help characterize very thin beds.

When a representative core database has been obtained, one can identify a representative set of *petrophysical bed types*, including both reservoir and non-reservoir beds, by clustering the core data into a set of narrow distributions. These bed types should be defined consistently with the geologic reservoir description. The core data are then analyzed to determine the distributions of the petrophysical properties for each bed type.

For comparison with log-analysis results, macroscopic averages of petrophysical properties over contiguous core intervals are computed using the averaging relationships discussed in Chapter 6. Bed-type fractions needed for these averages are obtained from digital core descriptions or photographs.

Terminal Splay / Distal Alluvial Fan,
Eocene,
Wilkins Peak Member of Green River Formation:
Sage Creek Canyon, Flaming Gorge, Wyoming, U.S.A.

Photo by Kevin Bohacs

8 Digital Core Imaging in Thinly Bedded Reservoirs

Overview

Introduction

Digital core photographic images are the highest resolution continuous datasets of reservoir properties that are available. The resolution can be as high as hundreds of samples per inch of depth, and beds as thin as a fraction of an inch can be easily resolved. Digital core images are derived from cores that have often been described in detail by geologists. Thus, they provide a link joining petrophysical analysis with geologic descriptions such as facies and depositional environment.

This chapter begins with a summary of basic concepts of digital imaging and continues with an introduction to digital core images and their acquisition, storage, and analysis. The potential applications of digital core images are numerous, but in our focus on estimating hydrocarbon pore-thickness in thinly bedded reservoirs we address only three aspects:

1) Classifying the bed type of each distinct region of the core image (e.g., as sandstone or shale, in the simplest case);
2) Quantifying the volume fractions of each bed type over the cored interval and/or selected subintervals;
3) Determining earth models for log convolution modeling and resistivity modeling (see Chapter 11).

Contents

DOI: 10.1306/1157791A13220

Digital Image Basics

Resolution

Resolution is a critical concept for the digital representation of core images. Resolution represents the level of detail contained in the digital image, and typically has units of pixels per length: dots per inch, points per inch, pixels per inch, dots per mm, etc. Figure 8.1 shows a 1-inch [2.5-cm] square segment of a core photograph, displayed at image resolutions of 100, 50, and 25 dots per inch (dpi).

Figure 8.1. Core photograph displayed at three resolutions.

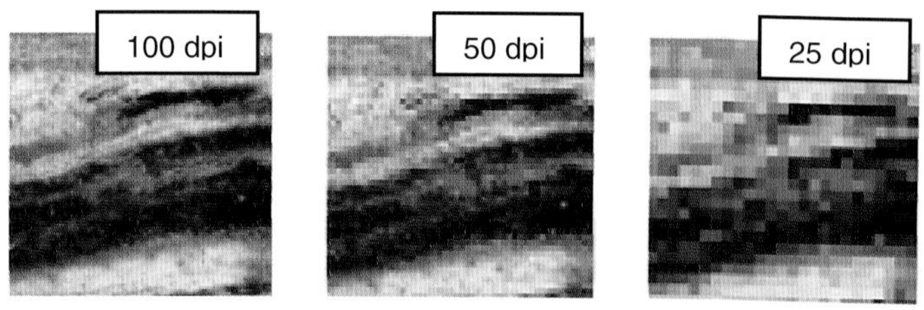

The ratio between the *image* resolution and *true* resolution is the *scale* of the image. The leftmost image of the core in Figure 8.1 has an *image* resolution of 100 dpi and a *true* resolution of 100 dpi. The size of a feature on Figure 8.1 is the same size as the feature on the actual core and the *scale* is 1:1, or true scale.

Resolution for scanned digital images is typically defined as the number of pixels per inch of the scanned object. If a 4-in. × 5-in. [10-cm × 13-cm] photograph were scanned into a computer at 200 dpi, the resulting image would be 800 pixels wide by 1000 pixels tall. The *image* resolution would be 200 dpi, regardless of the *scale* of the image. Its *true* resolution would vary depending on whether the photograph showed a mountain range or an amoeba (for example).

A general rule for setting the resolution of core photographs is to choose a value an order of magnitude greater than the smallest feature of interest. For example, if you are interested in 1.0-in. [2.5-cm] beds, the *true* resolution of the image should be at least 10 dpi. To resolve 0.1-in. [.25-cm] beds, the *true* resolution should be at least 100 dpi. The size of the digital image file is typically proportional to the square of the image resolution, so there is a tradeoff between acceptable resolution and acceptable file size.

The resolution of a scanned photograph is not always the *true* resolution of the final image. The resolution of the final digital image is equal to the scale of the photograph times the resolution of the scan. For example, if you scan half-scale core photographs at 400 dpi, the true resolution of the final image is 200 dpi.

Color

Digital color images consist of a two-dimensional array of data in which each pixel represents a color. Colors may be described digitally in several different standard formats. Each format requires at least three pieces of information. The most intuitive format is additive color, in which each color is described in terms of its proportions of red, green, and blue (*RGB color*). This is the format that corresponds to the method of display used by computer monitors.

Typically, the proportions of red, green, and blue are each recorded as one byte (8 bits with values 0 or 1). Each pixel has a red value between 0 and 255, a green value between 0 and 255, and a blue value between 0 and 255. Thus, each pixel in a RGB color image has 3 bytes (24 bits) associated with it. This corresponds to a number of colors equal to 2 to the 24th power, or roughly 16.7 million colors. Color stored in this format is sometimes called *24-bit color*.

Eight-bit color images are compressed versions of 24-bit images. Eight-bit images use lookup tables, also known as color maps, color tables, or palettes. These images are analogous to a 'paint-by-numbers' drawing in which each pixel is given an index number that references a position in the lookup table, where the actual RGB description of the color is stored. Color lookup tables typically contain 256 possible colors (8 bits). Eight-bit color is an efficient means of storing digital color images, requiring about one-third the file size compared to 24-bit color. However, for images with a large variety of colors, textures, shadows, or details, the quality of an 8-bit image can be poor. In recent years, 8-bit color has been made obsolete in favor of 24-bit and higher color resolution (e.g., 32-bit and higher); 24-bit remains the standard for nearly all core image applications.

The *color accuracy* of a single digital image, and the *color consistency* of a series of images, depends on many factors including lighting, film type and development, and subsequent digital processing. Methods of color calibration to ensure consistency in a series of core photographs are discussed below. Color consistency is a requirement for accurate digital analysis of the core image data.

File formats

Digital image file formats provide an agreed-upon set of rules for preparing, describing, and distributing image data in digital files. More than 100 such file formats exist. Fortunately, less than a dozen or so are commonly used.

It is outside the scope of this publication to discuss image file formats in detail. Here we simply note that TIF (tagged image file) and Kodak PhotoCD (PCD) are two widely used formats that are suitable for storing high-quality digital core images. TIF and PCD both support *24-bit color*. Two other very common formats are JPEG and GIF. GIF images should be used for display purposes only, not for quantitative analysis, as the format reduces color information. JPEG images can be used for digital analysis but it is important to realize information is lost in JPEG compression and artifacts may be inadvertently created.

Compression

Digital images require large amounts of disk space and memory. The images included in this chapter required tens of megabytes (MB) to store. The size of a digital core image database can approach hundreds of MB to as much as several gigabytes (GB). *Digital image compression* is a means to reduce the amount of space required to store digital images. There are numerous methods, or algorithms, for compressing digital image files. The *compression ratio* is the ratio of the size of the original data file divided by the size of the compressed data file.

Lossless compression occurs when a data file that has been compressed and then uncompressed and is identical to the original file. Lossless compression methods typically result in compression ratios on the order of 1.5 to 2. *Lossy* compression means that information is lost during the compression of a data file, so the original data cannot be recovered. For example, JPEG images use a Discrete Cosine Transform algorithm, which results in extremely high compression ratios (10 to 70) with significant loss of data. In general, lossy compression methods should be avoided unless images are being prepared purely for display purposes.

In creating digital core image databases, file compression may be considered when the available disk space is limited. Only lossless algorithms should be considered. The drawback of image compression is the inconvenience associated with decompressing images, and/or a slightly longer access time when opening compressed images in processing applications.

Introduction to Digital Core Images

Utility of digital core images

Digital core images are the highest resolution continuous datasets of reservoir properties that are available. The resolution can be as high as hundreds of samples per inch of depth, and beds as thin as a fraction of an inch are easily resolved. Digital core images are derived from cores that have often been described in detail by geologists. Thus, they provide a link joining petrophysical analysis with geologic descriptions such as facies and depositional environment.

In a thinly bedded reservoir it may be impossible to obtain an accurate depth tie between well logs and core plug data. However, borehole image logs and core photographs can often be integrated to obtain a detailed depth tie.

Finally, the techniques of digital image analysis can be applied to digital core images to provide quantitative information such as sand volumes, bed-type distributions, and bed-thickness distributions that might otherwise be obtained only through painstaking manual description and measurement of the core.

Methods of acquisition

Digital core images can be obtained in several ways:

1) Scanning core photographs directly,
2) Scanning negatives of core photographs,
3) Direct digital photography of the core,
4) Direct digital scanning of the core.

The first two methods begin with conventional film photography. Guidelines for photographing cores with conventional or digital cameras and guidelines for scanning conventional photographic prints are given below.

There are cost-effective commercial services that provide high-quality digital scanning of photographic negatives and write the scanned images to compact discs (CDs). One such service is the Kodak Photo CD. The standard Photo CD is designed for 35-mm photography and produces an image of 2048 pixels by 3072 pixels from a 24-mm x 35-mm negative. The Pro Photo CD is designed for larger film formats, or for higher-resolution scans of 35-mm negatives, and produces an image of 4096 pixels by 6144 pixels.

The Photo CD images are stored using a proprietary compression technology that allows up to 100 images to be written on a single CD. These PCD image files can be read by the standard photo-processing software packages, such as Adobe Photoshop® and Corel PhotoPaint®, and by Microsoft Office® applications.

Finally, there are specialized services that provide direct digital scanning of cores, bypassing conventional photography altogether. One such service is the Corimag^{TM} system, a product of Reservoir Laboratories AS (Trondheim, Norway). Another is the DMT Colour CoreScan® System, a product of Deutsche Montan Techologie GmbH (Essen, Germany). Information about these services may be found on the Internet at www.reslab.com and www.dmt-gmbh.net, respectively. We do not discuss these services further in this publication, except to note that increasing resolution and quality of digital images combined with increasing storage size of digital cameras will ultimately favor digital acquisition over conventional film.

Types of core images

Plain-light (or *white-light*) images are photographed under natural lighting conditions. The image is what one would see if one were standing above the slabbed core viewed under natural light. *UV-light* images are taken while the slabbed core is illuminated with an ultraviolet lamp.

Natural oil-bearing reservoir rocks often exhibit strong ultraviolet fluorescence. The oil absorbs the UV light and re-emits visible light. Typically, oil-bearing sands emit yellow colors; however, the brightness and color of the fluorescing oil is highly variable and depends upon the oil composition.

UV-light photography in oil-bearing rocks often enhances the contrast between reservoir beds and non-reservoir beds, making it easier to quantify the proportion of reservoir beds in a macroscopic cored interval. Figure 8.2 shows an example where the plain-light and UV-light photographs of a 1-ft [.3-m] core section are displayed side by side. Note the sharp contrast between shale and sandstone in the UV photograph on the right, compared to the limited contrast in the plain-light photograph on the left.

Figure 8.2. Plain-light (left) and UV-light (right) core photographs.

Following are a few cautionary remarks and recommendations associated with white-light and UV-light core images.

- *Plain-light*: Sand is not always brighter than shale. Do not apply quantitative analysis techniques to plain-light images before examining core carefully.
- *UV-light* (oil reservoir): Flushing can remove oil and reduce fluorescence, especially near the outer surface of the core. Compare the outer and inner portions of the image. If there is evidence of flushing, it may be necessary to crop an outer rind from the image before applying quantitative techniques.
- *UV-light* (oil reservoir): If the slabbed core was not well-preserved or was not photographed soon enough after slabbing, the sands may not fluoresce uniformly. Examine the state of the core and consider reimbibing the slabbed core.
- *UV-light* (oil reservoir): Reservoir sands do not always fluoresce. If the presence of non-fluorescing oil is suspected, compare core plug oil saturations with core images.
- *UV-light* (water or dry gas): Water and dry gas do not fluoresce. Use the plain-light image or consider re-imbibing the core with a synthetic fluorescing oil to obtain contrast. Re-imbibition may change the wettability of the core or contaminate the core with non-reservoir oils. Therefore re-imbibition should be applied only to the thin portion of the slabbed core, which is generally preserved for viewing only.

Core-to-log depth alignment

In thinly bedded reservoirs it may be impossible to obtain an accurate depth tie between cores and logs using only conventional log data and core plug data. However, an accurate tie may be critical when using core plug data to calibrate or validate a petrophysical model. Borehole image logs [Chapter 9] provide a high-resolution view of the borehole wall and, when data quality is sufficiently high, may allow a very detailed and accurate log-to-core depth-match with core image data. Figure 8.3 shows such a tie using an electrical borehole image (EBI) log on the left (shown in shades of orange; see Chapter 9) and plain-light and UV-light core images on the right. Beds as thin as two centimeters are aligned.

Core Photography Guidelines

The two keys to success

Two keys to successful core photography are:

1) Document the process;
2) Include as much additional information in the field of view of the photograph as possible: For example, scale cards, gray-scale cards, color cards, depth, well identification.

Slabbing the core

Before photography and plug sampling, the initially cylindrical core is *slabbed*: It is sliced parallel to the cylinder at a depth approximately one-fourth to one-third of the core diameter from the edge. The width of the slabbed face will be on the order of a few inches, depending on the diameter of the core and the depth of the slab. The core is cut at an angle 90° from the average strike of the beds to show maximum dip.

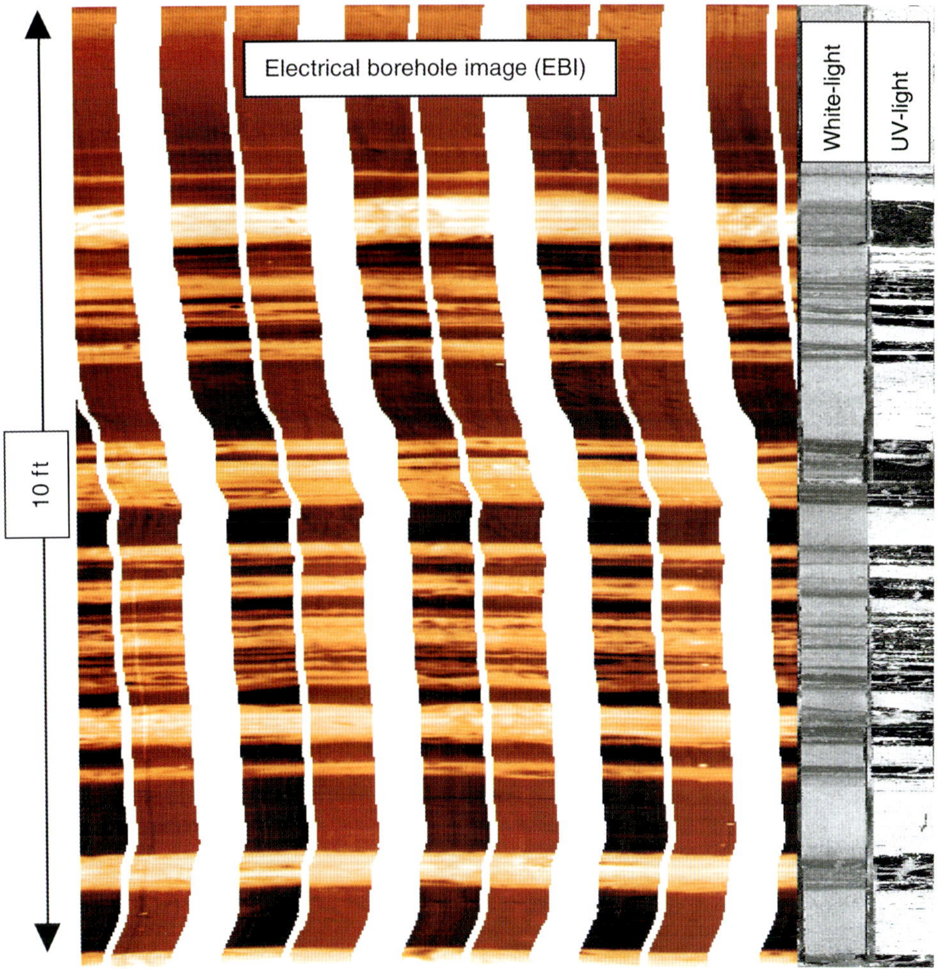

Figure 8.3. High-resolution depth tie between a borehole image log (left) and core photographs (right). Both white-light and UV-light core images are shown.

Figure 8.4 shows schematically the cross-section of a slabbed core. The deeper the slab, the wider will be the area exposed for photography. However, the depth of the slab limits the maximum length of core plugs cut from the main body of the core. There may be a trade-off between the requirement for longer core plugs and the need for maximum photographable width.

Table 8.1 shows the width of the slabbed core face as a function of the depth of the cut and the diameter of the core.

Table 8.1. Width of slabbed core face.

Diameter of core	Width of slabbed core (one-third cut)	Width of slabbed core (one-quarter cut)
D	0.943 D	0.866 D
2.5 in. [6.4 cm]	2.36 in. [5.99 cm]	2.17 in. [5.51 cm]
4 in. [10 cm]	3.77 in. [9.58 cm]	3.46 in. [8.79 cm]
6 in. [15 cm]	5.66 in. [14.38 cm]	5.20 in. [13.21 cm]

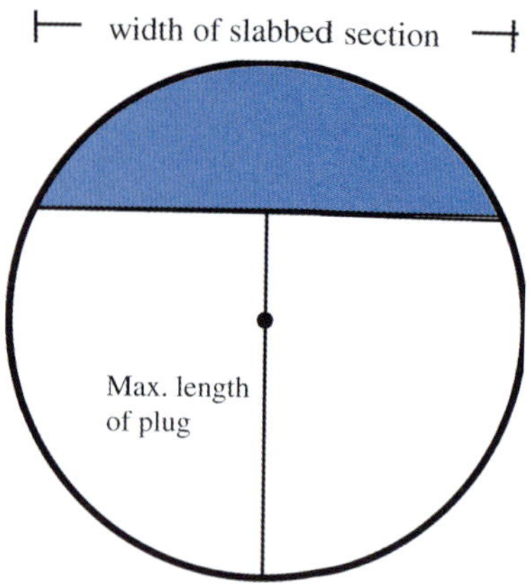

Figure 8.4. Schematic showing geometry for slabbing a core. The important elements include the diameter of the core, the width of the slabbed face, and the maximum possible length of a core plug.

Table 8.2 shows the maximum length of core plugs as a function of the depth of cut and the diameter of the core.

Table 8.2. Maximum length of core plugs.

Diameter of core	Maximum length of plug (one-third cut)	Maximum length of plug (one-fourth cut)
D	0.667 *D*	0.750 *D*
2.5 in. [6.4 cm]	1.67 in. [4.24 cm]	1.88 in. [4.78 cm]
4 in. [10 cm]	2.67 in. [6.78 cm]	3 in. [8 cm]
6 in. [15 cm]	4 in. [10 cm]	4.5 in. [11.4 cm]

Photographing the core

The process of core photography involves laying out the core so that the flat slabbed face lies horizontally face up. Photography is usually performed using the thin slab portion of the core. The core may be placed in a specially constructed cradle that will have a 120- or 35-mm camera and a bank of lights mounted directly above it. For half-scale core photography, individual foot-long core sections will be shot individually. For composite core photography, several sections of core will be photographed simultaneously.

Half-scale photography provides an ideal image resolution for digital analysis. With composite photographs the final resolution of the digital images may be too limited to resolve some small features.

Documenting core in field of view

The following items should be visible in the core photograph to aid in the identification and analysis of the digital images:

- The core section identification;
- An information card with the depth of the core section, name of the well, the company, *etc.*;
- Depth marks along the side of the core;
- A gray-scale card for checking brightness and contrast;
- A calibration card for checking color quality.

Figure 8.5 illustrates a well-documented core photograph shot at half-scale. Two of the 1-ft [2.5 cm] sections have been spliced together. Some of the identifying markers, including the original depth labels, have been digitally covered by opaque green rectangles to obscure the identity of the reservoir. Note the depth marks at 0.1-ft [.25-cm] intervals along the core. To ensure color consistency between sections of the white-light photograph, the gray-scale card and color card must appear in each photograph. The white-light photograph (center) and the UV-light photograph (top) were shot at different times and spliced together to form the final, complete half-scale image of each core section. In addition these photographs have been digitally annotated with core plug analyses at the correct depths.

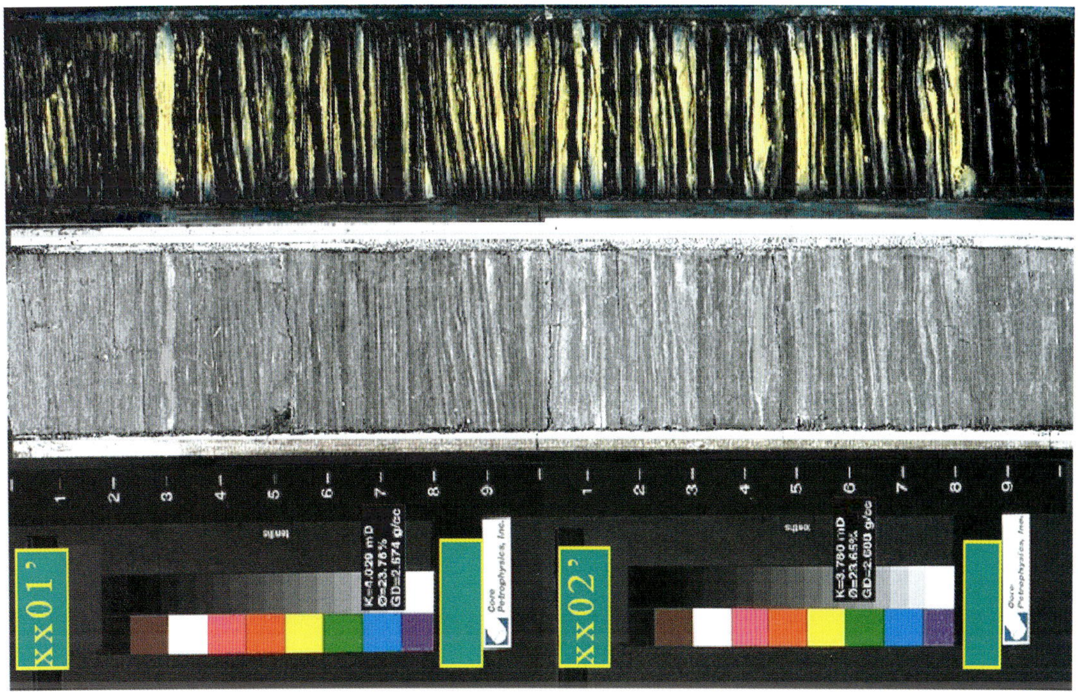

Figure 8.5. A well-documented core photograph.

Types of film used

Several sizes of negative are available for core photography: 35-mm, 70-mm, and 120-size. A 35-mm negative is 35 mm wide and 23.3 mm tall. A 70-mm negative is 70 mm wide and 46.7 mm tall. A 120-size negative is 2.25 in. square. Most core photography is done using 120-size film, which has higher resolution than 35-mm film, but is more expensive.

In terms of quantitative analysis, we see no significant difference between digital images acquired using 120-size negatives or 35-mm negatives. Either format is satisfactory. It is also significantly less expensive to generate Kodak PhotoCD data from 35-mm negatives than from 120-size negatives.

UV calibration standard

Calibrated photographs are essential to perform quantitative analyses. Whereas white-light photography typically includes a Kodak color card and gray scale card in the field of view, there is no standard counterpart for use in UV photography. The Kodak color cards do not fluoresce.

For calibrated UV photography, both a stable, consistent source of UV emission and a set of UV color calibration standards are needed. Figure 8.6 shows a UV image where hydrocarbon-imbibed polymethylmethacrylate UV standards were used. These are the same standards used in laboratory UV fluorescence spectrophotometers.

Figure 8.6. UV core image with a set of UV fluorescence standards in the field of view.

Scanning Core Photographs

Overview

The three critical concerns when scanning core photographs are that:

1) the scanning is consistent from one photograph to another;
2) the photograph is scanned at a sufficient resolution; and
3) the scanning reproduces the original colors and appearance of the core.

Figure 8.7 is an example menu from a typical scanning software package. Standard options include:

- the type of image (black-and-white, gray-scale, 8-bit color, 24-bit color),
- brightness and contrast, and
- scaling or resolution.

Color

The type of image selected should be *24-bit color* (millions of colors); 8-bit color is not satisfactory for image analysis and can create unwanted artifacts. Although most image analysis treats only the gray-scale data, it is recommended that the data be captured as 24-bit color images and subsequently converted to gray-scale and saved as separate files. The reason for this is two-fold: It allows for checking of color quality, and in some cases the color images can resolve information that cannot be resolved in the gray-scale data.

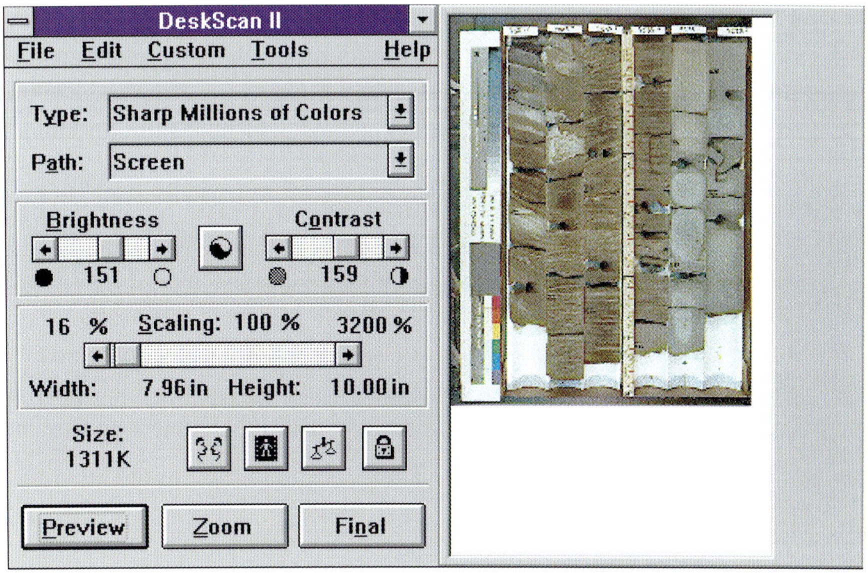

Figure 8.7. Typical flatbed scanning software menu.

Brightness

The brightness and contrast selected should be kept constant for all images scanned. Furthermore, the values should not be extreme because high contrast and brightness scans can introduce biases into the data.

Resolution

The scaling or resolution chosen is a compromise between resolution of small-scale features and available disk space. A general rule is to choose a *true resolution* that is an order of magnitude greater than the smallest feature of interest. For example, if you have true-scale photographs and are interested in 1.0-in. [2.5-cm] beds, the scan should be at least 10 dots per in. (dpi). If you are interested in 0.1-in. [.25-cm] beds, the scan should be at least 100 dpi.

The resolution of the scan is not always the *true resolution* of the final image. The resolution of the final image file is equal to the scale of the photograph times the resolution of the scan. For example, if you scan half-scale core photographs at 400 dpi, the resolution of the final image is 200 dpi.

File format

Scanned images should be stored in a file format that is portable among various platforms and software packages, and that retains all the information in the original image. Thus, if compression is used, it should be *lossless* compression. A format that has all these qualities is the 24-bit TIF.

Creating a Core Image Database

Overview

Once acquired, digital core image data must be stored in an organized form that facilitates analysis. We refer to this organized collection as a database even though its creation, and subsequent access to the data, does not necessarily require the use of formal database software. In this section we present a few general considerations for creating such a database, and describe the steps that were followed in a particular case study.

The core image database begins with a set of original image files that all share the same image resolution, file format, and documentation format. The set of images needs to be checked for color consistency, and calibrated if necessary, before further editing and analysis. For purposes of analysis each original image may be duplicated, edited, and stored in more than one format (e.g., both 24-bit color and gray-scale).

The final storage format chosen for the edited image file(s) that have been prepared for analysis will depend on the software package to be used in the analysis. For example, with standard photo-editing packages such as Adobe Photoshop®, it may be most expeditious to analyze each image in a separate file. However, with suitable specialized software, image analysis can be performed more efficiently when the series of core images is first spliced together into a single, large image file representing a complete contiguous core. For example, ExxonMobil uses a proprietary image analysis application.

Color consistency

There are two aspects to color quality: (1) how well a given digital core image matches the true color of the core, and (2) how consistent each core image is relative to others in the set. The first aspect involves *absolute* color calibration. The second aspect involves *relative* color calibration (*color consistency*). For most core image applications, good *color consistency* is sufficient; absolute color calibration requires considerable care during the processing of the data and also during acquisition of the photographs or images.

Color consistency is required to ensure that quantitative image analysis results are not biased by artifacts caused by variations in color, brightness and contrast between photographs.

A quick technique for checking color consistency is to examine the numeric RGB (red, green, blue) values for the color card shown in the field of view of each core photograph. Many digital imaging programs (e.g., Adobe Photoshop®) can display average RGB values for a selected area. As each image file is opened, several of the color patches in the calibration card should be selected and their RGB values noted. If the values are roughly consistent for all of the images, then the color consistency is good.

If significant variations are noted in the examination of the color patches, then calibration is required to achieve color consistency. Detailed steps for calibration depend on the specifics of software packages and will not be discussed here. However, one general approach to calibration would utilize the gray-scale card and focus on the following objectives.

1) Every point of the card should have equal RGB values on all photos.
2) A given point on the card should have consistent RGB values from one photograph to the next.
3) The white end of the card should have RGB = (255, 255, 255).
4) The black end of the card should have RGB = (0, 0, 0).

Database example

The process of database construction will be illustrated with a case study from the Gulf of Mexico. Three wells were cored through a reservoir interval. Core lengths ranged from 40 ft to 71 ft [12 m to 22 m] per well.

High-quality, half-scale, plain-light, and UV-light core photographs were taken for each of the three wells. In the UV-light photographs, oil-bearing reservoir sands appear yellow-orange, and non-reservoir mudstones appear dark or black. An example scan of one half-scale photograph is presented in Figure 8.8.

The first step in the core image preparation process was the scanning of the raw half-scale photographs. These photographs were originally scanned in 2-ft [.6-m] increments from laminated prints. The base images were scanned at a resolution of 400 dpi (for a *true resolution* of 200 dpi) in 24-bit color and stored in TIF format. Uncompressed file sizes for the 2-ft [.6-m] images were approximately 36–40 megabytes (MB) in size. Therefore, the original dataset for a 50-ft [15-m] core would be about one gigabyte (GB) in size.

After scanning, individual UV-light images were cropped from the photographs in 1- to 2-ft [.3- to .6-m] sections. Slight rotation of some photographs was necessary prior to this step. The cropped images were inspected for evidence of mud filtrate invasion near the edges of the core, and the images were further cropped. This process reduced the 24-bit image size to approximately four MB per 2-ft [.6-m] section, or 100 MB for a 50-ft [15-m] core.

Figure 8.9 illustrates this cropping and several subsequent processing steps that are described below.

The cropped, 24-bit color images were converted to 8-bit gray-scale for analysis. This conversion further reduced the size of the image datasets. The resulting images were approximately 2 MB per 2-ft [.6-m] section, or 50 MB for a 50-ft [15-m] core.

For some core sections, an additional image-processing step was necessitated by the presence of missing and/or "rubbled" core. Such intervals were systematically masked (i.e., converted to null values) from the cropped 24-bit images before conversion to gray-scale. Fine depth adjustments of these core images were performed to ensure the preservation of a true depth scale in the vertical dimension. Null values are color-coded in green on the gray-scale image in Figure 8.9.

Figure 8.8. Example half-scale plain- and UV-light scan. Image file is stored in TIF format and is approximately 38 MB in size (uncompressed).

Note in Figure 8.9 that the print fold at approximately 12160.8 ft [3706.6 m] is deleted from the 24-bit "masked" image before conversion to gray-scale, and the image sections are joined to correct the depth scale (each tick mark on the original image represents 0.1 ft [.03 m]).

Regions in the image field where there is no useful data, such as the folds and missing core segments discussed above, must be represented in the image files in a unique way: As nulls (missing values) or as a unique color not found in the actual image data. Otherwise, if these regions become confused with white or black parts of the actual photograph, errors in analysis will result.

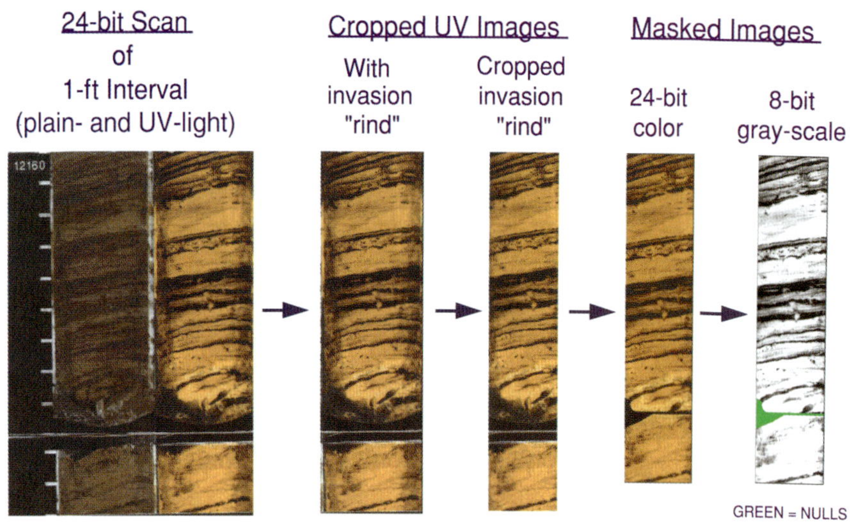

Figure 8.9. Graphical illustration of core image preparation process.

Finally (not illustrated in Figure 8.9), the 1- or 2-ft [.3- or .6-m] image sections, prepared as described above, were concatenated into a single continuous image file for each continuous core. The concatenated image file was used as the basis for digital analysis to determine the reservoir sand fraction in each core. This analysis is described in the next section.

Analysis of Digital Core Image Data

Overview

For the purpose of estimating hydrocarbon pore-thickness (HPT) in a thinly bedded reservoir, the analysis of digital core image data has two main steps:

1) Classifying the bed-type of each distinct region of the core image (e.g., as sandstone or shale, in the simplest case);

2) Quantifying the volume fractions of each bed type over the cored interval and/or selected subintervals.

The first step may often be accomplished by digitally manipulating the image into a form where each bed type is completely distinct — e.g., white for sandstone and black for shale. The two examples below illustrate this case, where a brightness threshold can be applied to the image (UV or plain-light) to separate sandstones from shales. In other cases the discrimination of bed types may be more difficult and may require a subjective process of manually drawing the outline of each bed, then assigning it a distinct color to represent its bed type.

When each bed has been classified, the quantification of bed-type fractions is simply a matter of counting the number of pixels within each bed type and dividing by the total number of non-null pixels in the image.

How each of these steps is accomplished in practice depends on the particular software used for image analysis. The examples below illustrate the use of Adobe Photoshop®.

Note that we are not aiming here to estimate the *producible* sand fraction. A thin sand with a high core-plug permeability and high hydrocarbon saturation may have low effective permeability because of bedding geometry beyond the scale of the borehole. The issue of hydrocarbon producibility must be addressed by other methods which are not discussed here.

Well 1: Sand fraction from UV threshold

The first example is based on the core image database described above. These cores represent a deep-water Gulf of Mexico oil reservoir interval which consists of a sharp-based, fining-upward succession of thinly bedded sandstones and mudstones that were deposited by low density turbidity currents in unchannelized, distal distributary lobes. The interval is characterized by a high net sand volume ratio but low resistivity due primarily to thin-bed effects. It is evident in Figures 8.8 and 8.9 that the sandstone and shale beds can be clearly distinguished in the UV photographs of the oil-saturated core. The quantitative discrimination of sandstone and shale is accomplished by applying a threshold to the gray-scale UV photographs. Figure 8.10 illustrates this procedure using Adobe Photoshop®.

The uppermost image in Figure 8.10 is a 2-ft [.6-m] section of gray-scale photograph, prepared as described above. The center image is the thresholding menu showing the histogram of the gray-scale. The threshold level can be set numerically or by sliding the triangular pointer at the base of the histogram. The lower image is the "thresholded" result, where points below the gray-level threshold are now all black, and those above the threshold are all white. The histogram tool in Adobe Photoshop® can now be applied to this lower image to determine the proportions of sandstone (white) and shale (black).

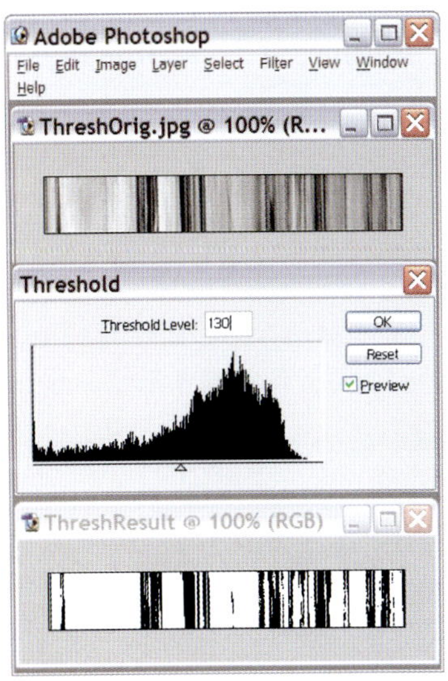

Figure 8.10. Thresholding in Adobe Photoshop® (Well 1).

Well 2: Sand fraction from plain-light threshold

This example comes from another deep-water Gulf of Mexico reservoir comprising unconsolidated sands of Pliocene age. Six dominant facies are present: massive sands, classic turbidites, sandy debris flows, muddy debris flows, mud-clast conglomerates, and pelagic mudstones. Massive sands and pelagic mudstones comprise about 85% of the reservoir interval, so most sections of the core image were either 100% net or 0% net. In the remaining sections, the image data were used to quantify the net sand fraction.

Figure 8.11. Applying threshold to plain-light image.

The example shown in Figure 8.11 comes from the gas cap, where the UV photographs are relatively featureless. There is little contrast in the original plain-light photograph (left), but the sands and shales are distinct in the contrast-enhanced image (center), making it easy to select a brightness threshold. The thresholded image (right) is then used to calculate the sand fraction.

Summary

Digital core images may be acquired by scanning standard high-resolution photographs, or by direct digital photography or scanning of the slabbed core. Core images should be digitized at a resolution at least ten times the smallest feature that needs to be resolved. For example, to resolve beds 0.1-in. [.25-cm] thick, a resolution of at least 100 dots per inch (dpi) should be used. Digital core images should be stored in 24-bit color using only lossless compression schemes and a file format that maintains high image quality.

Cores should be photographed in both white and ultraviolet (UV) light; UV-light photography often enhances the contrast between non-reservoir and oil-bearing reservoir beds. Digital core images and borehole image logs can often be integrated to obtain a high-resolution depth match between cores and logs.

Two keys to successful core photography are, first, to document the process; and second, to include as much identifying information as possible in each photograph's field of view: For example, scale cards, gray-scale cards, color cards, core depth, and well identification. Gray-scale and color cards are necessary to assure color consistency between photographs. Either 120-size or 35-mm film provides adequate resolution for core photography.

Once acquired, digital core image data must be stored in an organized form that facilitates analysis. The core image database begins with a set of original image files that all share the same image resolution, file format, and documentation format. The set of images needs to be checked for color consistency, and calibrated if necessary. For some applications the individual core images need to be concatenated into a single image file representing the complete continuous core.

For the purpose of estimating hydrocarbon pore-thickness (HPT) in a thinly bedded reservoir, the analysis of digital core image data has two main steps: First, classifying the bed-type of each distinct region of the core image; and second, quantifying the volume fractions of each bed-type over selected intervals. In the simplest case, the first step is accomplished by a digital thresholding technique that leaves one bed type white and all the others black. Then the fraction of the white bed type is quantified by digitally counting the white pixels and dividing by the total number of pixels in the core image. In an oil reservoir, thresholding is most often applied to the UV image.

Fluvial / Lacustrine,
Eocene,
Wasatch Formation:
Canon Pintado, Colorado, U.S.A.

Photo by Kevin Bohacs

9 Borehole Image Logs in Thinly Bedded Reservoirs

Overview

Introduction

In recent decades, advances in digital technology have enabled the transmission of ever-increasing volumes of data uphole from both wireline logging and logging-while-drilling instruments, and increasing numbers of these tools now are capable of producing images of one kind or another. This chapter begins with a summary of the variety of image-capable logging tools that exist at the time of writing. Then the focus is narrowed to the electrical micro-imaging tools, which have the most general applicability for the evaluation of thinly bedded reservoirs. These tools provide the highest-resolution wireline data available for resolving beds down to 1 or 2 in. [2.5 or 5 cm] in thickness.

In this chapter and elsewhere in the text we use a non-standard name and acronym, Electrical Borehole Image (EBI), to refer generically to the class of electrical micro-imaging tools without specifying a particular service company's device.

Since the late 1980s, the petrophysical literature is replete with examples of EBI data being used for the estimation of sand volumes in thinly bedded reservoirs. However, in most cases the resolution and accuracy of these estimations have not been rigorously tested.

Our conservative approach to the application of the EBI tools is outlined here. Careful application of EBI image data requires some understanding of tool principles and data processing methods. These topics are summarized below, followed by a discussion of specific applications in thinly bedded reservoirs, with examples. The examples emphasize both the value of the EBI tools and the pitfalls that can lead the unwary to questionable results.

Contents

DOI: 10.1306/1157792A13220

Introduction to Borehole Image Logs

Types of borehole image logs

Over the past two decades, increasing numbers of logging tools have been engineered to produce images of one kind or another. Advances in digital technology enable the uphole transmission of ever-increasing volumes of data from both wireline logging and logging-while-drilling (LWD) instruments. Multiple sensors or rotating sensors are used to generate azimuthal images.

Table 9.1 summarizes the kinds of image-producing logging tools that exist at the time of writing, with reference to their utility for the evaluation of thin beds. The focus in this chapter is on electrical micro-imaging tools, which can produce high-resolution images showing detailed bedding features. Acoustic micro-imagers are also discussed briefly below.

Table 9.1. Logging tools capable of producing images.

Log Type	Description
Wireline electrical micro-image (EBI)	Arrays of small contact buttons generate images of the borehole wall, resolving features less than 1 in. [2.5 cm] in size. These logs are the main focus of this chapter.
Wireline acoustic micro-image	A rotating acoustic transducer measures the amplitude and transit time of echoes from the borehole wall. Potentially useful for thin beds.
Wireline focused-resistivity array	Enhanced versions of the classic laterolog tool, equipped with azimuthal contacts that produce cylindrical images with vertical resolution near 1 ft [.3 m].
Wireline "array" induction	Enhanced versions of the classic induction tool, with multiple transmitter-receiver spacings for multiple depths of investigation. Data can be displayed as a resistivity image in the plane of the borehole and radial directions.
LWD focused-resistivity array	Produces rudimentary azimuthal image of resistivity near the bit while drilling. Useful for real-time information, not considered here for detailed thin-bed evaluation.
LWD circumferential density image	Rotating source and detectors produce a rudimentary azimuthal image of bulk density while drilling.

Electrical micro-imaging tools (EBI)

This chapter focuses on the electrical micro-imaging tools (EBI) and their applications in thin-bed evaluation. Each major logging service company deploys one or more versions of these tools. The current versions at the time of writing are summarized in Table 9.2.

Table 9.2. Electrical micro-imaging tools.

Service company	Tool name and description
Baker Atlas	STAR Imager^{SM} (Simultaneous Acoustic and Resistivity Imager). Combines electrical and acoustic micro-imagers in a single tool. Electrical imager is designed for water-based mud.
Baker Atlas	EARTH Imager^{SM}. Combines electrical and acoustic micro-imagers in a single tool. Electrical imager is designed for non-conductive mud.
Halliburton	XRMI^{TM} (X-Tended Range Micro-Imager). Electrical micro-imager for water-based mud.
Schlumberger	FMI^{TM} (Formation Micro-Imager). Electrical micro-imager for water-based mud. Schlumberger was the originator of the electrical micro-imaging tool.
Schlumberger	OBMI^{TM} (Oil-Base Micro-Imager). A version designed to work in oil-based mud.

The EBI tool design is an elaboration of the classical dipmeter tool. It employs arrays of small button electrodes (on the order of 5 mm [.2 in.] in diameter) arranged on pads at the ends of several articulated arms that push the pads into contact with the borehole wall. Currents pass through these buttons, into the formation, and back to the tool via a remote return electrode. The individual button currents are affected by small-scale variations in the resistivity of the invaded zone. These currents are sampled several times per inch, producing a dense array of data that is processed uphole to generate an image of the borehole wall.

As an example of the physical design of the EBI tools, Figure 9.1 shows a schematic of the Schlumberger FMI^{TM} (Formation Micro-Imager, a mark of Schlumberger). This tool has four articulated arms with conductive pads attached. Each pad has two rows of current buttons, electrically insulated from the pads, plus a hinged flap that carries two more rows of buttons offset azimuthally from those on the pad. Altogether the FMI^{TM} has 192 buttons and produces an image that covers 80% of the circumference of an 8.5-in. [21.6-cm] borehole.

Figure 9.2 shows a 10-ft [3-m] section of a micro-resistivity image obtained with the Baker Atlas STAR Imager^{SM} service. The STAR Imager has six arms and a total of 144 buttons. The micro-resistivity image is interrupted by six near-vertical stripes representing regions of the borehole that were not contacted by one of the tool's six pads. The lateral drift of these stripes corresponds to the slight rotation of the tool as it is pulled uphole. Darker regions on the image represent lower resistivity and brighter regions represent higher resistivity. The sinusoidal shapes that extend across the top and center of the image result from the dip of bedding features relative to the borehole.

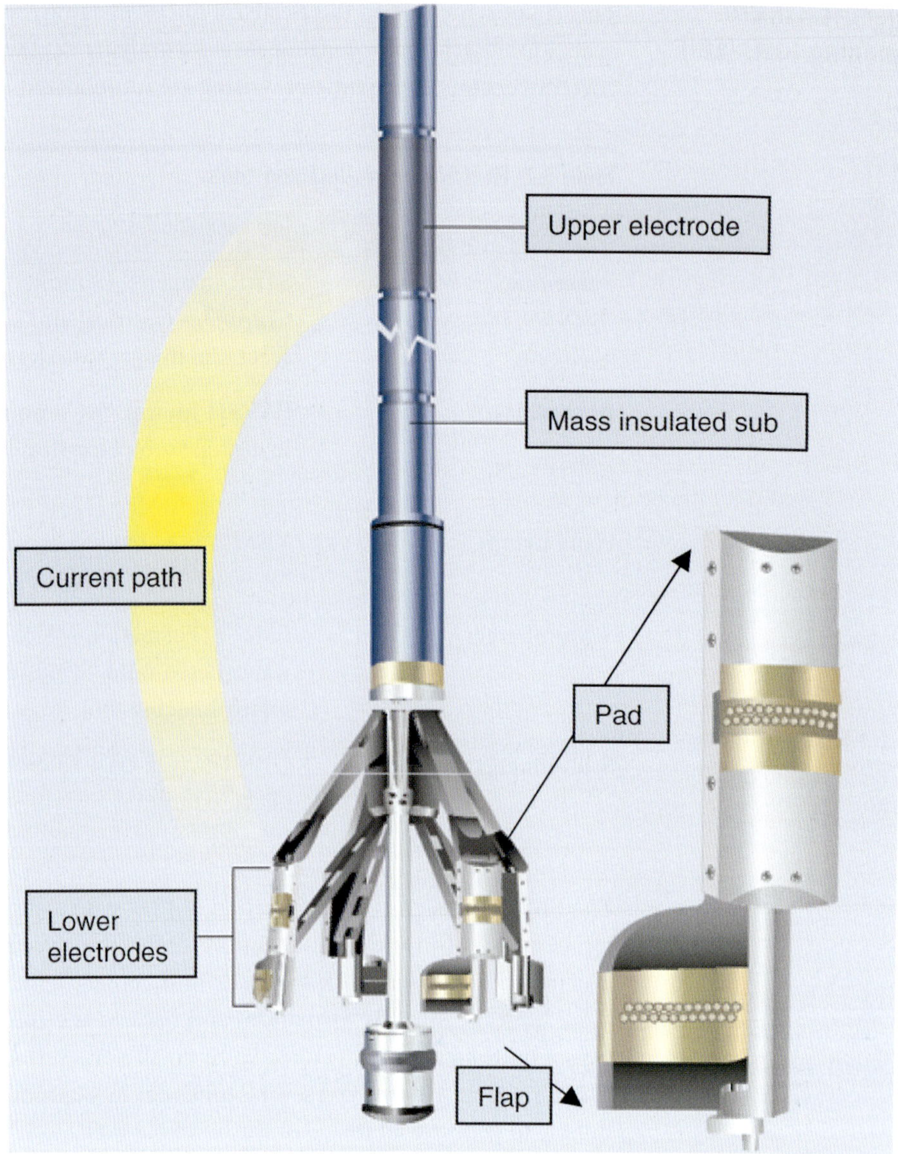

Figure 9.1. Schematic of Schlumberger FMI™ tool showing contact buttons on pads and flaps, and illustrating path of current to upper return electrode (figure courtesy of Schlumberger).

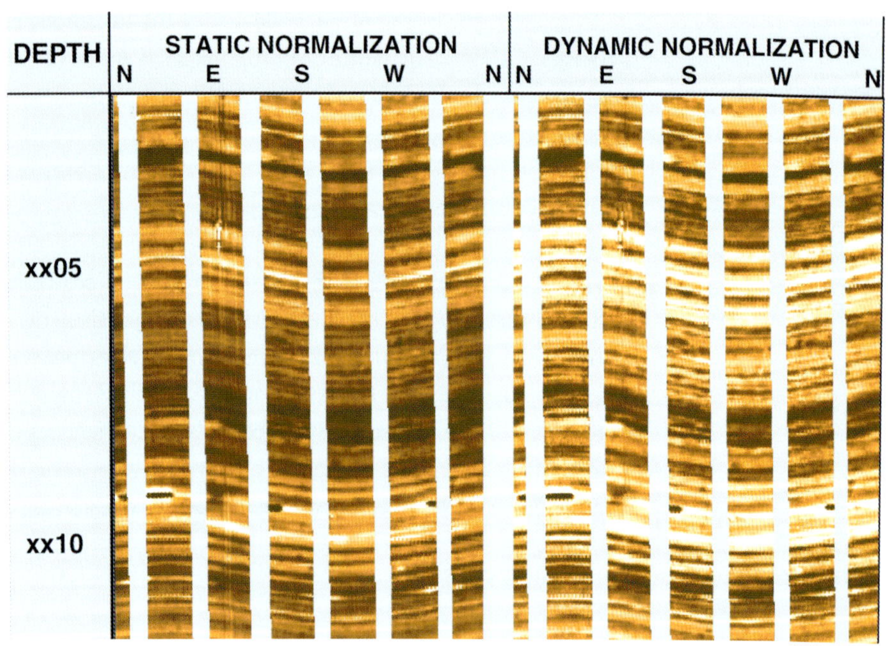

Figure 9.2. Borehole images produced by STAR Imager. Statically normalized micro-resistivity image is on the left and dynamically normalized micro-resistivity image is on the right (courtesy of Baker Atlas).

Acoustic micro-imaging tools

Acoustic micro-imaging tools use a rotating transducer, centered in the borehole, to send and receive acoustic pulses that echo off the borehole wall. The travel time and the amplitude of the returning echoes are recorded, and these measurements are used to produce two separate images of the borehole wall. The names for the service companies' variants of the acoustic micro-imaging tool are summarized in Table 9.3.

Table 9.3. Acoustic micro-imaging tools.

Service company	Tool name and description
Baker Atlas	CBIL℠ (Circumferential Borehole Imaging Log). Stand-alone acoustic imaging tool.
Baker Atlas	STAR Imager℠ (Simultaneous Acoustic and Resistivity Imager). Combines electrical and acoustic micro-imagers in a single tool.
Halliburton	CAST-V™ acoustic imaging tool.
Schlumberger	UBI™ (Ultrasonic Borehole Imager, a mark of Schlumberger).

Figure 9.3 is a schematic illustration of the Schlumberger UBI™ turned on its side, showing the rotating transducer at the bottom (right) and the in-line centralizer that maintains the position of the transducer relative to the borehole wall.

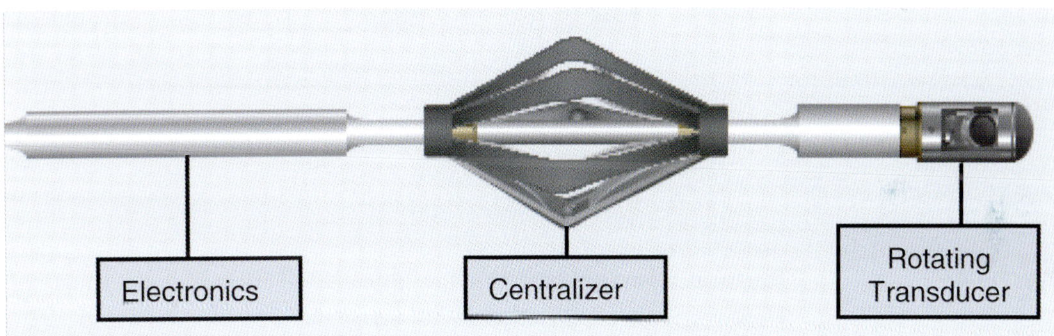

Figure 9.3. Schlumberger UBI™ acoustic micro-imager (figure courtesy of Schlumberger).

Figure 9.4 shows a 10-ft [3-m] section of images from the Baker Atlas CBIL℠ service. The left track is the amplitude image and the right track is the travel-time image. The sinusoidal shape of the fractures seen in these images results from their dip relative to the borehole.

The amplitude and travel time of the acoustic echoes respond mainly to the shape and texture of the borehole wall and to the relative acoustic impedance of the geological features imaged. Acoustic images are particularly useful for the analysis of faults, fractures, borehole stability and geometry, and the in-situ stresses that affect these features. They are also used for casing inspection. Acoustic images tend not to show as many fine bedding features as micro-resistivity images, which are generally more useful for the applications addressed in this publication.

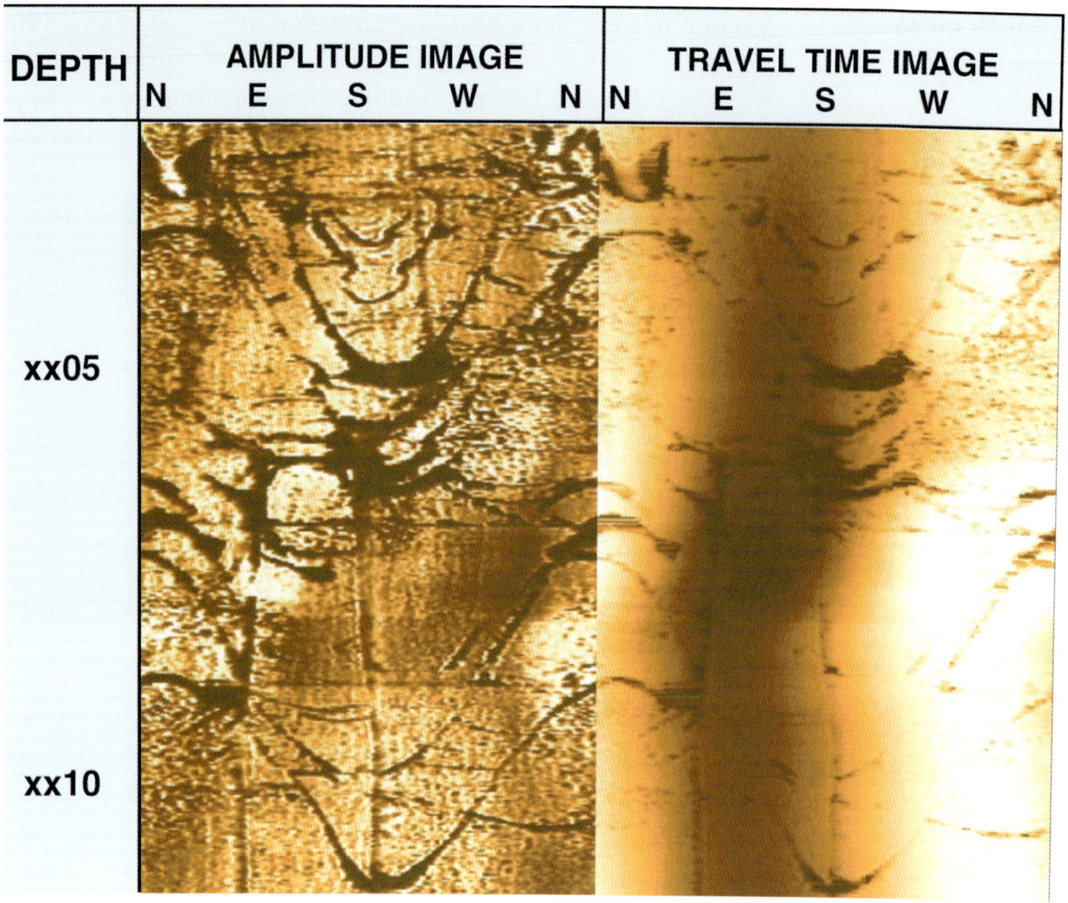

Figure 9.4. Borehole images produced by CBIL acoustic imaging service (courtesy of Baker Atlas).

Principles of Electrical Micro-imaging

Overview

Electrical micro-imaging logs are potentially very useful in evaluating thinly bedded reservoirs. However, proper application of EBI image data requires some understanding of tool principles and data processing methods. In this section, the measurement principles of the EBI tools are introduced as follows:

- the basic physics of the tool's voltages and currents and the environmental conditions affecting individual button measurements are discussed;
- the variable electrode voltage on the EBI tool and its impact on individual button measurements are explained;
- the principles of depth (or speed) corrections are introduced; and
- the requirements for accurate azimuthal orientation are reviewed.

The basic principles of each EBI tool are the same, regardless of service company and specific tool generation. Most of the information summarized below has been obtained in reference to the Schlumberger FMI™ tool.

Basic physics

EBI measurements are similar to older dipmeter tool measurements. Specifically, an electrical potential difference (a voltage, V_e) is applied between the lower portion of the tool (the pads, arms, and buttons) and a return electrode further up the tool string. Refer to Figures 9.1 and 9.5. This voltage difference causes current to flow out of the higher potential pads and buttons, through the formation and borehole, and to the lower potential return electrode. Each button current is proportional to the local conductivity of the formation and can be described by Ohm's Law as in Equation 9.1.

$$i_b = k\, V_e\, C \tag{9.1}$$

where i_b is the button current, C is the local formation conductivity, and k is a factor that may vary slowly with depth, related to the geometrical factor of the tool and the current path between the button and the return electrode.

The current emitted from each individual button is monitored and the array of current values is displayed as an image, as illustrated in Figure 9.2. By convention, light image colors (e.g., yellow) are associated with relatively resistive features in the formation (i.e., low current density), and dark colors (e.g., brown) are associated with relatively conductive features in the formation (i.e., high current density).

Figure 9.5 illustrates the "passive focusing" that forces individual button currents into the formation in a direction approximately normal to the borehole wall. The conductive pads are maintained at the same potential as the buttons but are insulated from them. The current field emitted from the pads produces the passive focusing effect on the independent button currents.

Electrical micro-image button measurements are typically uncalibrated. In other words, two runs of two different EBI tools over the same portion of a borehole may yield similar-looking processed images, but the absolute value of the current measurements may be significantly different. This factor limits well-to-well quantitative comparisons of image data.

At regular time intervals, the current from each imaging button is measured (e.g., 192 current values every 0.02 s for an FMI™ logged at 1500 ft/hr [457 m/hr]). The resulting array of currents yields a high-resolution, electrical "map" of the borehole wall when converted to depth and properly oriented.

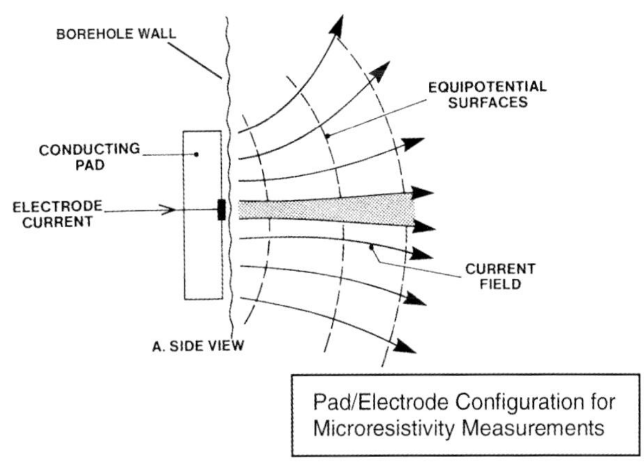

Figure 9.5. Schematic current field for an EBI tool pad. Modified from Ekstrom et al., 1986.

Vertical resolution and detection limit

The vertical resolution of an EBI tool cannot be less than the diameter of its individual buttons — a few tenths of an inch. However, as with other logging tools, various edge and environmental effects degrade the resolution of the measurement to something greater than the button diameter. In an empirical comparison with limited core data, Trouiller et al. [1989] estimated a 2-in. (5-cm) vertical resolution for Schlumberger's older-generation 4-pad FMS™ tool. Beyond the inherent resolution of the tool design, environmental effects discussed below can "defocus" the measurement, degrade the image resolution, and sometimes bias the location of bed boundaries.

Formation features that are smaller than the button diameter may be detected by an EBI tool. The limit of detection depends in part on the local conductivity contrast associated with the feature. For example, open fractures with apertures in the sub-millimeter range are commonly detected by these tools because of the high conductivity contrast between the mud-filled fracture and the resistive host rock.

Environmental effects

Electrical micro-image button currents and the resulting images are affected by various physical and electrical conditions that exist in the logging environment. Here we summarize several of these conditions which often complicate the interpretation of EBI images. More details can be found in Trouiller et al. [1989] and Ekstrom et al. [1986].

1) Pad contact. As pads lose contact with the borehole, the equipotential lines of the current field shift progressively into the mud column resulting in "current defocusing" and diminished image resolution. This effect is most pronounced near pad edges and for highly conductive mud (low R_m) or high R_{xo}/R_m ratios. Decentralization of the tool in deviated wells can result in poor pad contact with the upper surface of the hole.

2) Mudcake. Current focusing decreases with increasing mudcake thickness. Also, in the case of thin porous oil sand and shale laminations, "mudcake overlap" from sand beds onto adjacent shales may result in the resistive sand beds appearing thicker than their true thickness.

3) Invasion. The EBI depth of investigation is shallow and variable; it often does not extend deeper than the invaded zone. When the resistivity of the invaded zone (R_{xo}) approaches the resistivity of the surrounding shales (R_{sh}), defocusing and loss of bed resolution can result even when there are no problems with pad contact or mudcake.

4) Drifting conductivity baseline. The sensitivity of the button currents is affected by geological and/or environmental factors that influence the current path in the interval between the pads and far return electrode, several m uphole. These factors can contribute to a slow drift of the "conductivity" baseline with depth. Thus, for any given applied voltage, EBI apparent conductivity values may differ for beds or intervals with the same true conductivity, and such beds may therefore appear with differing colors in the EBI image.

Tool potential ("EMEX")

During logging, the EBI tool's applied potential (V_e) could theoretically be held constant and then button current would be proportional to formation conductivity as in Equation 9.1. However, to maintain an adequate signal response and comparable tool sensitivity in all formations, V_e is often varied slowly during logging. When the tool senses an overall increase in resistivity between the pads and the return electrode, V_e is increased to force more current into the formation. Conversely, when the tool senses a decrease in formation resistivity, V_e is decreased to lessen the current.

With the Schlumberger FMI™ tool, V_e is referred to as the "EMEX" voltage. During logging the EMEX voltage is recorded every 0.1 in. [.25 cm]. An "EMEX correction" is usually performed during post-processing of FMI™ data to remove the effects of the variable voltage on the raw current measurements. In this way, FMI™ image values are generated that are more proportional to true formation conductivity. If this correction is not applied, a more conductive formation may appear "brighter" than more resistive formations in the logged interval. Figure 9.6 is an example from the Gulf of Mexico where the shale-rich section in the bottom (left) quarter of the logged interval appears "brighter" than the resistive, thinly bedded sand-rich section in the upper (right) half of the interval before the EMEX correction is applied.

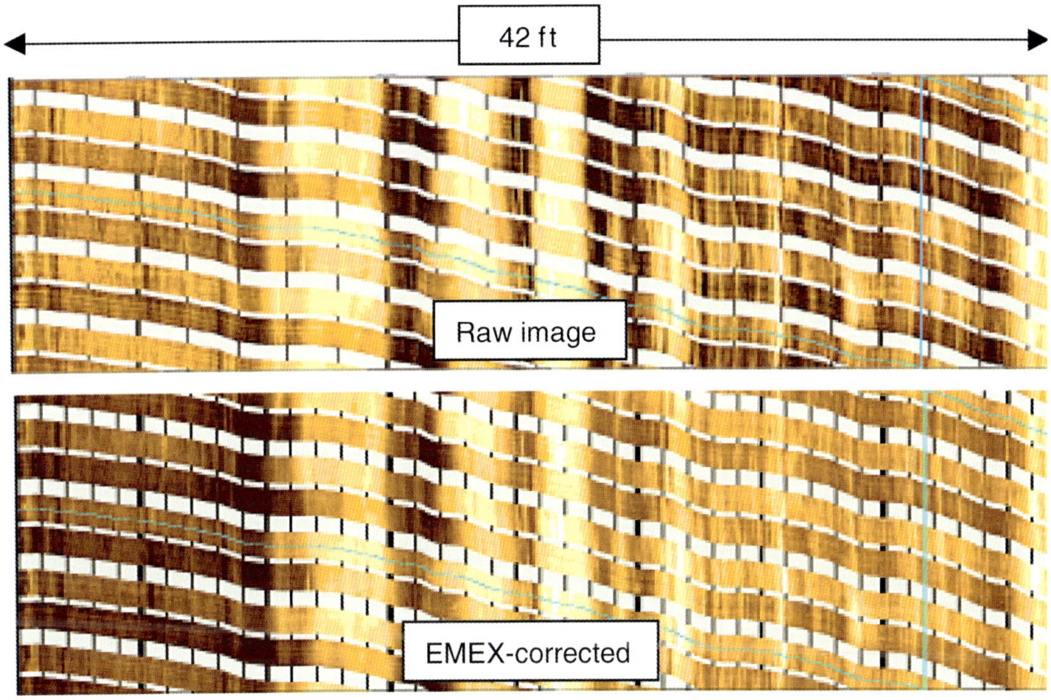

Figure 9.6. Raw and EMEX-corrected FMI™ images.

Depth registration

To produce a coherent and accurate image, the data from the EBI tool buttons must be registered accurately, both in terms of their positions relative to each other and with respect to depth. Exact down-hole positioning of wireline logging tools is a complex task because data are acquired at regular *time* intervals rather than *depth* intervals, and because friction can cause irregular motion and varying speed as logging tools are pulled uphole. Moreover, with the EBI tools, there is the additional complication that, at any given time, the data from each row of buttons are coming from different depths.

"Speed corrections" are data post-processing methods whereby the relative depth positions of EBI buttons are corrected for instantaneous differences in tool and cable velocity. Modern EBI tools incorporate downhole accelerometers and magnetometers to provide the positioning data necessary for these speed corrections.

Numerous published examples comparing image logs with core suggest that EBI button currents can be placed with remarkable accuracy at substantial reservoir depths. However, in spite of many advances in log depth registration, the depth of EBI measurements remains an imperfect estimate, and the accuracy of this estimate will vary between fields, wells, and even intervals of the same well. Nonetheless, it is important to note that speed-corrected EBI data probably are the most accurately depth-registered log data in the well.

Tool and image orientation

Electrical micro-images must be oriented with respect to a known reference, such as true north. An estimate of the three-dimensional orientation of a specific pad or button at any given time requires knowledge of three sets of information:

1) the geometric location of the pad/button relative to a reference pad/button;
2) the magnetic declination and inclination of the earth's magnetic field;
3) borehole size.

Electrical micro-image tools incorporate downhole magnetometers to orient the tool with respect to the earth's magnetic field, and the tool's arms are used to provide caliper measurements.

EBI Data Processing

Overview

Data processing is critical for optimal application of EBI measurements. In fact, the images generated from the raw EBI current data are themselves a product of post-acquisition data processing.

In a typical EBI processing stream, the raw button currents are transformed using a variety of algorithms in an appropriate sequence. In some cases, these algorithms numerically alter data (e.g., static normalization), and in other cases the algorithms geometrically alter data (e.g., accelerometer-based speed correction).

The operating company that acquires EBI data may have the processing done entirely by the logging service company, or may process the data in-house using commercial or proprietary software.

This section provides a brief introduction to the main steps in EBI data processing. Discussion of specific processing software packages, or more detailed processing guidelines, is outside the scope of this publication.

Processing flow chart

Figure 9.7 is a generalized flow chart of an EBI processing stream which is applicable to any EBI tool and any borehole image processing software package. The initial step labeled "Data collection, loading, reformatting" is highly specific to the logging vendor and the processing software and will not be discussed here, except to note that raw EBI data are not initially stored as an image, but as a set of "log" curves, one for each button current. Dip processing is also outside the scope of this publication. The various data corrections, static and dynamic image normalization, and calibration of the EBI "conductivity" are summarized below.

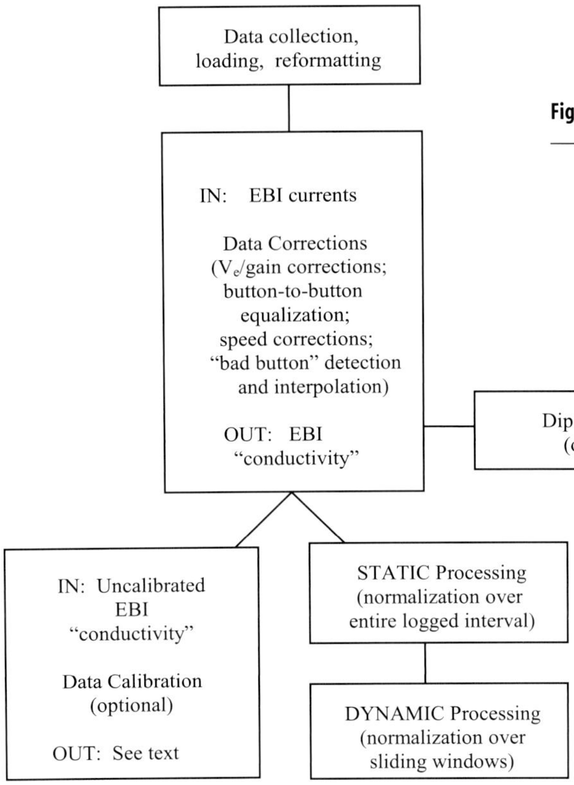

Figure 9.7. Generalized EBI data-processing flowchart.

Data corrections

A sequence of data corrections is listed below with brief descriptions. Several of these were introduced in the previous section. Corrections are performed in the order listed.

1) Voltage correction. To maintain an adequate signal response and comparable tool sensitivity in all formations, the applied voltage (V_e in Equation 9.1) may vary during logging. This voltage is recorded with the same frequency as the button currents. Correcting the button currents for this variable source voltage is a standard processing step and is recommended; see Figure 9.6 for an example.

2) Button equalization. The raw or voltage-corrected values of two or more buttons measuring a formation or event of constant resistivity can and often do exhibit different values. Several borehole effects that contribute to this phenomenon were previously described (e.g., differences in pad contact and/or mudcake thickness). The so-called button equalization process was developed to address this problem. Figure 9.8 illustrates the effect of button equalization on an FMI™ log. In the uncorrected image on the left, pad and flap segments are in pairs with the flap to the right. The flap images are notably darker. In the corrected image on the right, pad and flap images have similar brightness and contrast.

   In most cases, button equalizaton is *recommended*. However, in some applications where quantitative use is made of the image data, it may be preferable to use only the pad or pads that are in good contact with the borehole since poor contact reduces the precision of the image. An example of such an application is the estimation of bed boundaries for use in resistivity modeling.

3) **Speed correction.** Speed correction adjusts the relative depth positions of the EBI buttons for instantaneous differences in tool and cable velocity. In general, more accurate depth registration will be obtained by this process and speed correction is *recommended.* Depth estimates from datasets that lack speed correction can be significantly in error, and speed variation artifacts can be severe. This is especially true for logging runs with numerous pulls.

   Figure 9.9 shows the same FMI™ segment as Figure 9.8, before and after speed correction. Note the improved continuity of the thin beds in the speed-corrected image on the right. In this example there is only minor stretching and shrinking, but in cases of high differential tool and cable motion, the amount of depth shifting may be significantly greater.

4) **Bad-button interpolation.** During the computation of button-equalization statistics (see above), a given button can exhibit a mean and variance that is unreasonably different from adjacent buttons. If so, the button is considered faulty or "dead," and a new value is determined using linear interpolation of neighboring values.

   Faulty and/or dead buttons are very distracting on final image displays (they appear as constant value or null events that parallel the pad edge). Therefore, it is generally *recommended* that the bad-button correction be applied. However, it is important to recognize that this process creates data, and that the user cannot determine how much data have been created. Therefore, it may be useful to also generate an uncorrected image across a short interval for comparison.

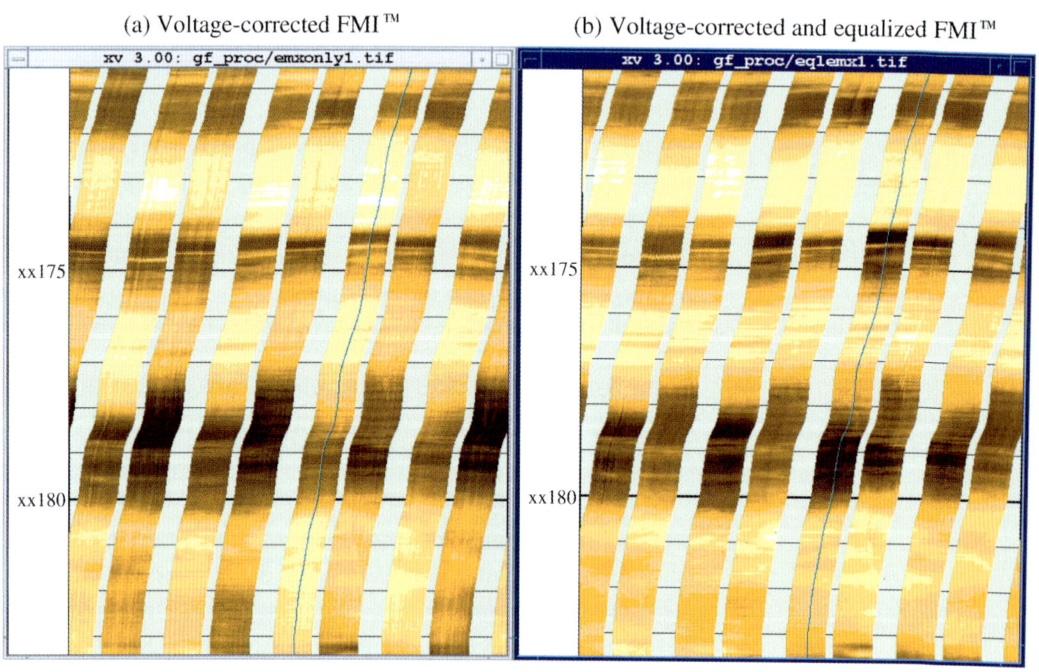

Figure 9.8. FMI™ image before (left) and after (right) button equalization.

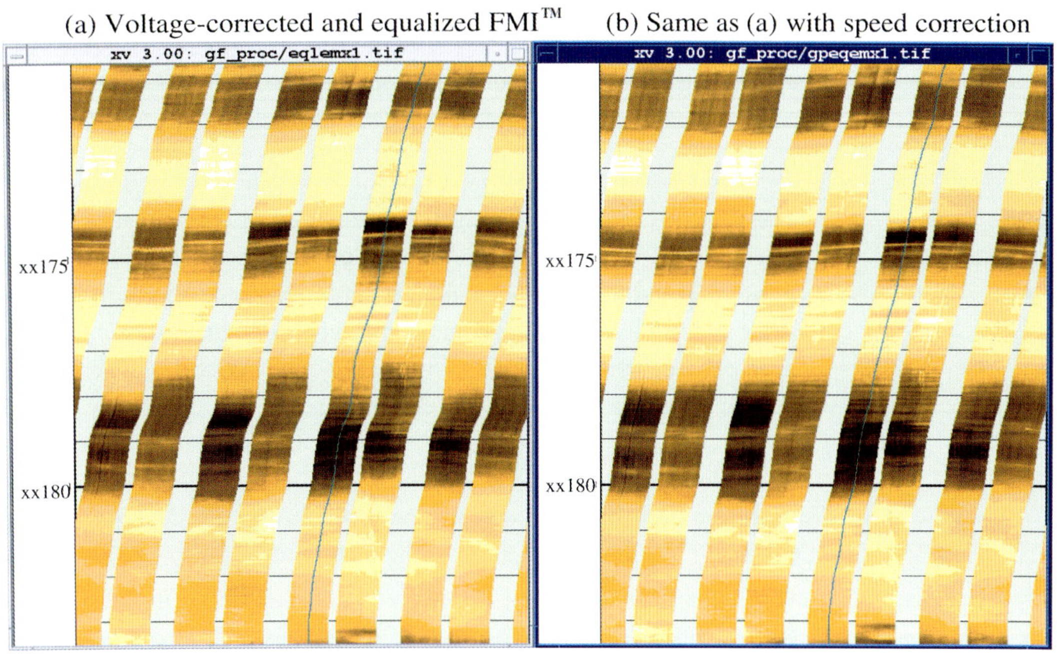

Figure 9.9. FMI™ image before (left) and after (right) speed correction.

Resistivity "calibration"

As was discussed previously, the EBI button data are fundamentally different from other resistivity devices in that (a) the tools utilize passive rather than active current focusing, and (b) the button measurements typically are uncalibrated. In other words, the EBI button arrays are not designed to measure either true or invaded formation resistivity.

To address this limitation, schemes have been devised which utilize shallow-reading, calibrated resistivity devices (e.g., a shallow laterolog) to "scale" the button data to more robust measurements of invaded formation resistivity. Specifically, for selected calibration zones, linear relationships are determined between vertically averaged EBI data (x) and shallow resistivity data (y) to apply an appropriate gain and offset to the EBI data. In this way, the high-resolution information from an EBI is transformed to a low-resolution base-line that approximates the calibrated shallow resistivity value.

Details of these calibration schemes are not discussed here. We offer below a few comments about their potential utility for the typical quantitative applications of EBI data in thinly bedded reservoirs.

A typical quantitative EBI application is to select a conductivity "cutoff" or "threshold" that yields a net or gross sand volume consistent with core-derived sand volume. This threshold is then applied in uncored intervals to estimate sand volume. In this application, a single linear remapping of EBI data will not affect the outcome of a threshold-based quantitative analysis. Data that exceeded a given threshold prior to scaling will simply exceed a different threshold after scaling. *Thus, resistivity "calibration" is not generally useful for this thresholding application.*

In many resistivity modeling applications, bed boundaries are determined from squared EBI data. These squared data are typically scaled to produce a first-pass "resistivity" that is input to the modeling application and subsequently modified. *The calibration schemes discussed here provide a convenient means of generating scaled "resistivity" data that can be used in these modeling applications.*

Image normalization

The primary purpose of EBI image normalization is to enhance contrast for subsequent display, whether it be on a computer monitor or paper print. The two most common schemes are referred to as *static* and *dynamic* normalization. The former process produces less local contrast than the latter and results in an image that is more proportional to formation conductivity. Dynamic normalization enhances local variations in conductivity and is generally used for feature recognition (e.g., crossbed enhancement, bed definition).

The following discussion focuses on the method of "histogram equalization" utilized by Schlumberger for both static and dynamic normalization of FMI™ images. Histogram equalization transforms the "conductivity" values of the FMI™ array to 8-bit color or gray-scale values such that all colors or gray-levels occur with equal frequency in the output image (that is, the initial distribution is transformed to an approximately uniform distribution). See Figure 9.10.

With ***static*** normalization, the histogram equalization process is run over the entire logged interval or a user-specified zone. The image data histogram is partitioned into fewer color bins such that each bin contains approximately the same amount of data. The user may select 1 to 64 colors for the normalization. A value of 1 would result in a monotone display; a value of 2 would yield a binary image.

In a ***static***-normalized image, the relative resistivity scale is partially preserved across the processed interval. A given color from one portion of the image represents a similar (although not necessarily *the same*) "resistivity" value as the same color elsewhere on the image. As the number of color bins decreases from 64 to 2, the range of original "resistivity" values represented by a given color increases. An example static image is presented for an FMI™ interval in Figure 9.11.

Like static normalization, the ***dynamic*** normalization process is run across a window of user-specified length. In this process, however, the window is typically small (e.g., 2 ft [.6 m]) and moving (or

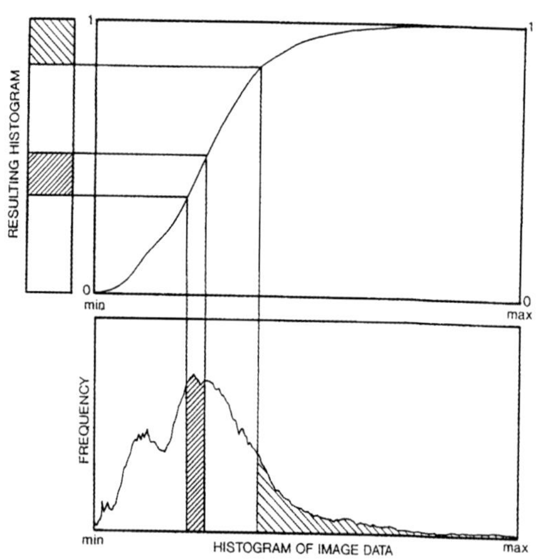

Figure 9.10. FMI™ histogram equalization. Image current or "conductivity" values are apportioned among several color or gray-scale bins such that each bin contains roughly the same volume of data (represented by the area under the initial frequency curve at base of figure). (Modified after Serra [1989].)

"dynamic"). Specifically, the equalization process is performed in a sliding window of user-specified length with 75% overlap between windows (i.e., the window slides 25% of its length for each equalization).

Dynamic normalization enhances the contrast of an image such that local features are more recognizable. It is useful for highlighting bedding, fractures, textures, and sedimentary structures. Unlike static normalization, the process does not preserve the relative resistivity scale of the data. The same color from two segments of the image does not necessarily represent the same or even similar resistivity values. An example dynamic image is presented in comparison to a static image of the same interval in Figure 9.11.

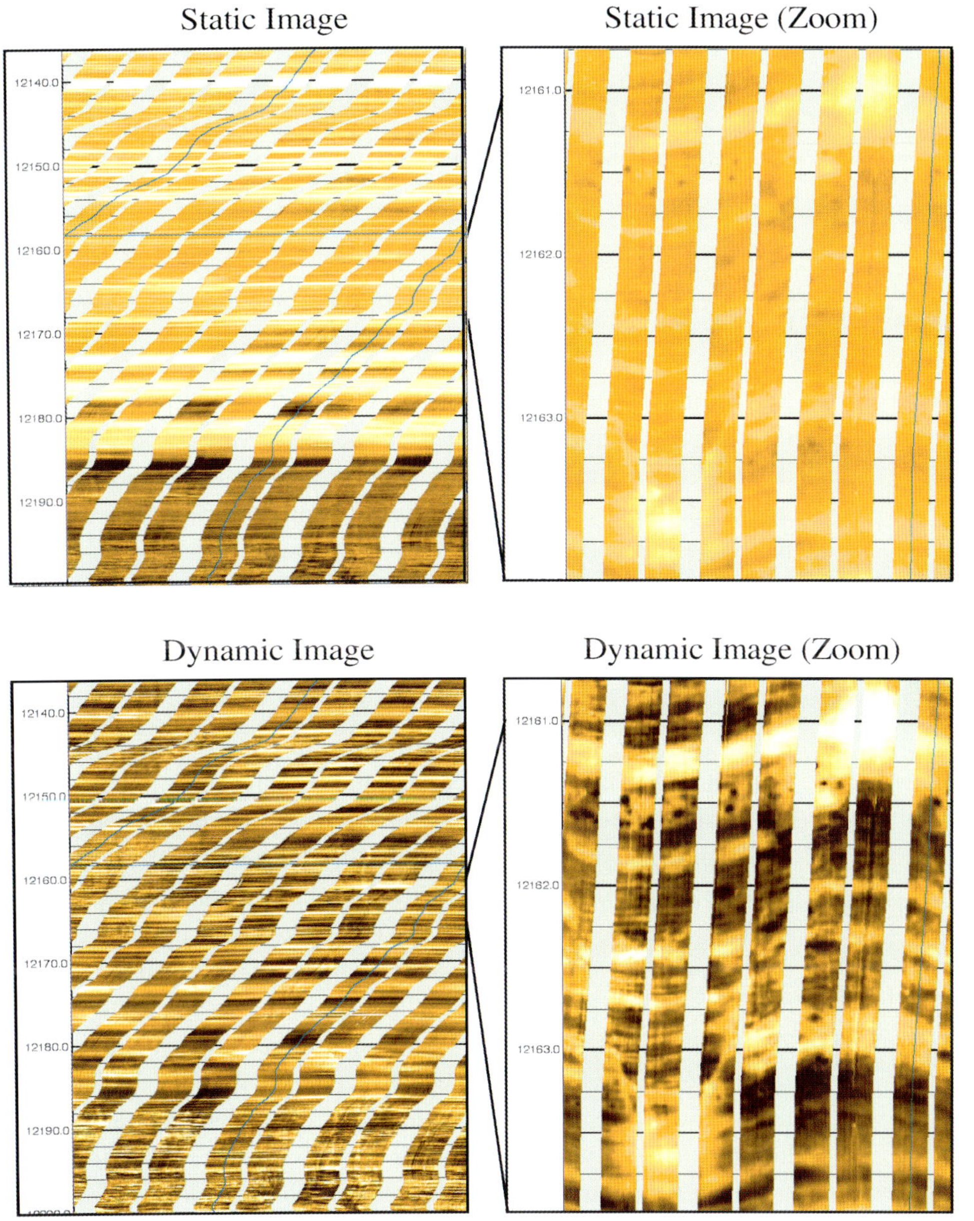

Figure 9.11. Static- and dynamic-processed FMI™ images for thinly bedded turbidite section from the Gulf of Mexico. Sand-rich interval above 12185 ft [3714 m] is identified in the static image (upper left). Dynamic processing highlights local resistivity contrasts (e.g., fracture at 12163 ft [3707 m]; resistivity "halo" at trough of fracture; possible conductive pyrite at 12161.5 ft [3706.8 m]).

Recommendations for the application of static and dynamic normalization are summarized below.

- Both *static* and *dynamic* normalization are recommended for all EBI image data that the user wishes to display graphically. The partially preserved EBI "resistivity" scale of the *static* image is useful for recognizing gross changes in facies and lithology. The high-contrast *dynamic* display is useful for recognizing local bedding characteristics, structures, and textures.
- Static normalization is not recommended for most semi-quantitative applications because the processing step alters the original information contained within the image. For example, in a threshold analysis of a static image over a short interval, errors can increase with a decreasing number of colors (or bins).
- Dynamic normalization is not recommended for semi-quantitative applications where a preserved resistivity scale is desired. However, this normalization scheme may be useful for certain semi-quantitative applications, such as fracture or structural orientation analyses, that are outside the scope of this publication.

EBI Image Artifacts

Common image artifacts

Some common EBI image artifacts and possible explanations are provided in Table 9.4. Some of these are processing artifacts that may be improved by adjusting appropriate parameters; others are acquisition artifacts that cannot be corrected. Note that most of the artifacts are best identified at expanded vertical scales (e.g., 1:5). For additional information on these and other artifacts, the reader is referred to Bourke [1989].

Thin-bed EBI Applications: Sand Fraction

Overview of method

Since the late 1980s, the petrophysical literature is replete with examples of borehole image data being used for the estimation of sand volumes in thinly bedded reservoirs. For examples see Boyeldieu and Jefferys [1988], Bourke [1989], Buscher [1989], Serra [1989], Troullier et al. [1989], Adams et al. [1990], Luthi [1990], McNaboe [1991], Pol and Fekete [1993], Serra et al. [1993], Sovich and Newberry [1993], Fam et al. [1995], Ramamoorthy et al. [1995]. In most cases the resolution and accuracy of these estimates have not been rigorously tested.

The standard EBI "sand count" technique involves selecting a conductivity *threshold* such that all depths with suitably-processed EBI "conductivity" above the threshold are classified as non-reservoir, while depths with EBI "conductivity" below the threshold are classified as sand. In the literature, this threshold technique is typically considered applicable above the gas- or oil-water transition zone, and it assumes that hydrocarbon-bearing sands will appear bright (or resistive) and shales will appear dark (or conductive) on the EBI images. Furthermore, if all the sands exceed suitable porosity and permeability cutoffs, the gross sand volume from threshold analysis may be considered equal to the net sand or net pay volume.

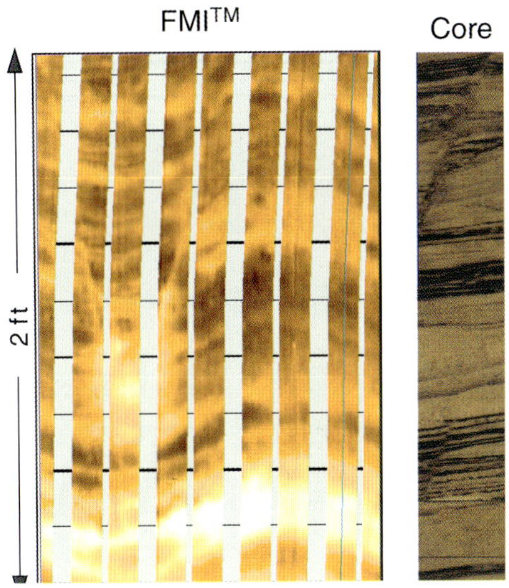

Figure 9.14. EBI and core images across a 2-ft [.6-m] interval.

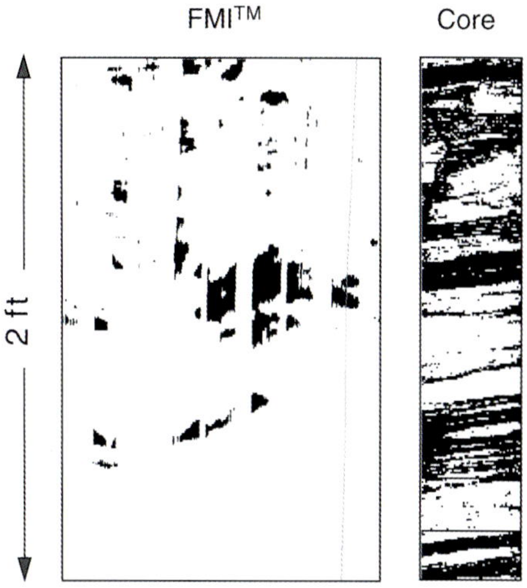

Figure 9.15. Thresholded EBI and core images across a 2-ft [.6-m] interval.

The lack of detailed correspondence seen in Figure 9.15, which is characteristic of much of the cored interval, shows that in this well the EBI thresholding technique cannot be used to obtain detailed bed boundaries and net-sand estimates even after calibration to 30 ft [9 m] of high-quality core image data.

Example 2: Reversed EBI image

Figure 9.16 is an example first introduced in Chapter 7. The figure shows a very detailed and precise correlation between an EBI image and the corresponding core image across a 10-ft [3-m] thinly bedded reservoir interval. Note, however, that the EBI image is reversed: it is bright in the shales and dark in the sands, where the sands and shales are clearly identified in the ultraviolet core image. This reversal is the result of two factors. The shales have very low porosity and thus are relatively resistive, and the sands have been flushed by saline drilling mud and are relatively conductive.

Figure 9.17 shows the thresholded UV image (bottom) and a thresholded and color-inverted image from a single pad of the EBI data (top). The two datasets were thresholded independently; that is, the EBI threshold was not calibrated to match the core image. The EBI image in Figure 9.17 has been color-inverted: Dark regions are shales and light regions are sands, corresponding to the UV image. Note the very close correspondence of bed boundaries in the two images.

The sand fractions computed independently from these two images are as follows: Core image = 65% sand, EBI image = 59% sand. The core image appears to have a few brightness artifacts within the shales that make the core-based sand fraction estimate higher despite the close correspondence of bed boundaries in the two images.

Note that, if we did not have the core image for comparison, the EBI image (Figure 9.16) might been interpreted in the standard way, with dark beds interpreted as shales and bright beds interpreted as sands. In this case the apparent sand fraction would have been 41% instead of 59% and the distribution of sand versus shale would have been completely reversed!

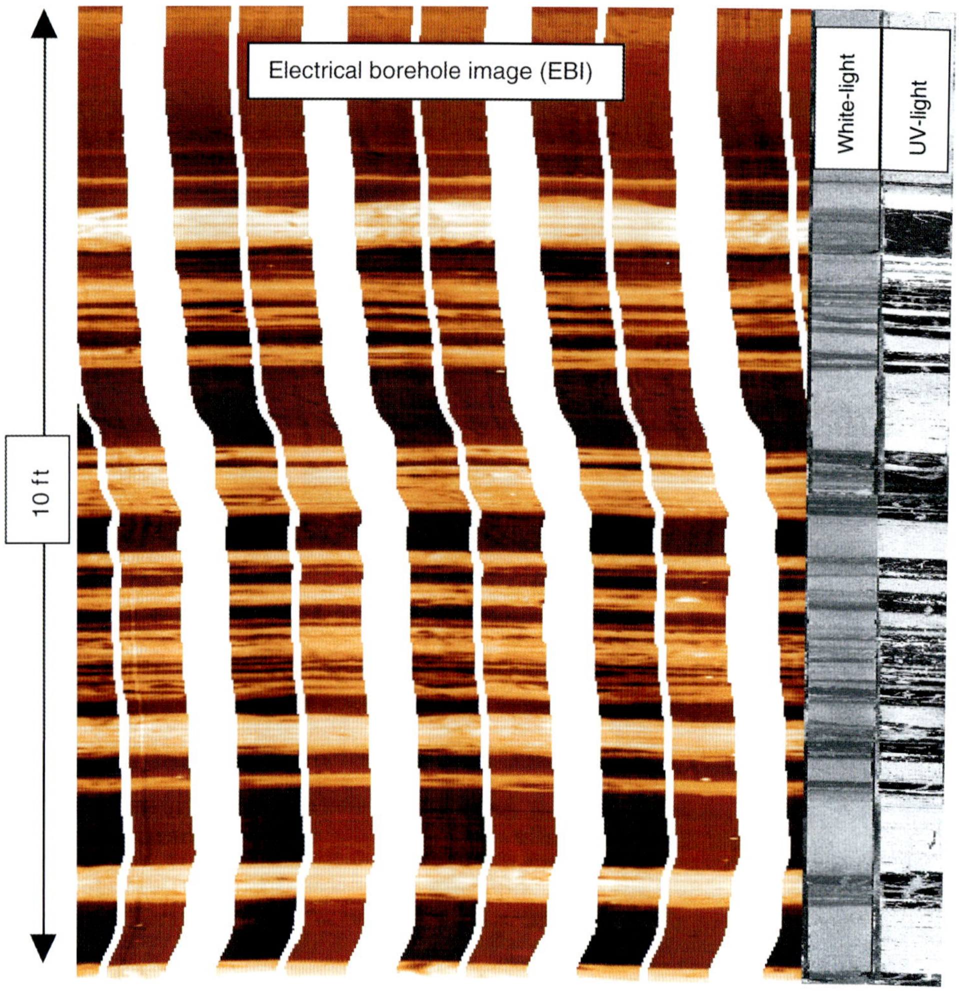

Figure 9.16. High-resolution depth tie between a borehole image log (left) and core photographs (right, both white-light and UV-light).

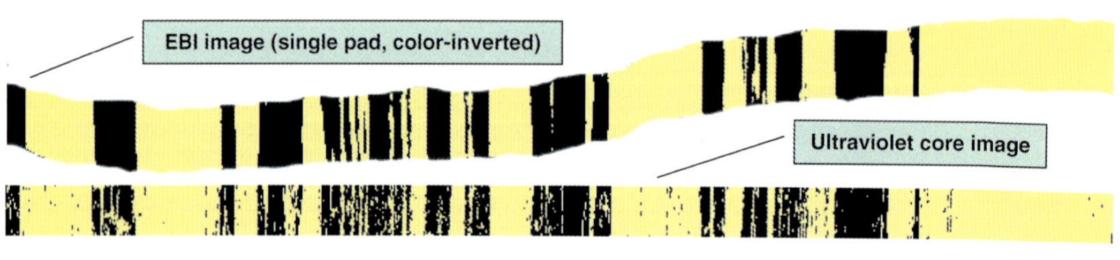

Figure 9.17. Thresholded UV core image (bottom) and independently thresholded EBI image (top).

Summary

We use a non-standard name and acronym, Electrical Borehole Image (EBI), to refer generically to the class of electrical borehole micro-imaging tools. Each major service company offers its own version of this tool. EBI logs provide the highest-resolution wireline data available for distinguishing beds down to 1 or 2 in. [2.5 or 5 cm] in thickness.

EBI measurements are similar to older dipmeter tool measurements, but they incorporate a much larger number of button electrodes on 4 or more pads that are pressed against the borehole wall. The current through each button is proportional to the local conductivity of the formation. The processed button currents are displayed as a colored image. The button data are not generally calibrated as quantitative conductivity or resistivity measurements.

The upper limit for the vertical resolution of an EBI log is determined by the diameter of the button electrodes, typically a few tenths of an inch. In practice, environmental factors such as mudcake, poor pad contact, and invasion often limit the actual resolution to 1 or 2 in. [2.5 or 5 cm] and may also distort the locations of bed boundaries.

A complex series of data-processing steps is required to convert raw EBI button currents into a coherent borehole image. Initial steps include correction for varying electrode voltage; correction for varying speed and orientation of the tool and its pads; and equalization of button responses. For display, the data are normalized to maximize image contrast. "Static" normalization produces an image in which similar brightness levels corresponds roughly to similar conductivities in different parts of the borehole, while "dynamic" normalization maximizes local image contrast without regard to a consistent relationship between brightness and conductivity.

The main applications of EBI data in thin-bed evaluation are the location of bed boundaries and the determination of net sand fraction. The standard EBI "sand count" technique involves setting a conductivity *threshold* to discriminate reservoir from non-reservoir. The preferred approach is to calibrate this threshold by comparison to core image data before it is applied to uncored intervals. Even with this cautious approach, the reliability of the extrapolated threshold analysis varies widely from one case to another.

Deep-water (Basin Floor Turbidites), Carboniferous, Ross Formation: Kilcher Cliffs, Loophead Peninsula, Ireland

Photo by Rob Wellner

10 Modeling Log Responses in Thinly Bedded Reservoirs

Overview

Introduction

Well-log measurements cannot be used directly to determine hydrocarbon pore-thickness in thinly bedded reservoirs because of the limits imposed by the vertical resolution of the logging tools [Chapter 1]. To overcome these limits we use mathematical models that describe how logging tools respond to thinly bedded formations. In fact, such modeling is the foundation of our approach to thin-bed evaluation. These mathematical models range in complexity from simple approximate equations to detailed numeric simulations that require high levels of computing power.

This chapter introduces the concepts and terminology of log response modeling: earth models, forward and inverse modeling, the types of forward models and the methods of inversion that are useful in thin-bed analysis. Specific models for lithology, porosity, and resistivity logs are also described.

The modeling techniques introduced here are the building blocks for the analysis methods described in later chapters — the methods that make up our integrated approach to thin-bed evaluation. In particular, the practice of resistivity log modeling (forward and inverse) is discussed in some detail in Chapter 11. The application of convolution modeling is discussed in Chapter 11. The application of Monte Carlo inversion with statistical earth models and interval-average log response models is discussed in Chapter 12.

Contents

DOI: 10.1306/1157793A13220

Earth Models

Definition

Any mathematical model of a logging tool's measurement requires, as a primary input, a geometric and petrophysical description of the earth region that is the *support* for the measurement. This description, which we call an *earth model*, consists of

1) A subdivision of the earth region into smaller regions, each of which can be described by a fixed set of petrophysical property values; and

2) The sets of petrophysical property values that characterize each of the sub-regions of the earth model.

The simplest earth model has only a single homogeneous region with constant petrophysical properties. Such a model can be termed *zero-dimensional* (0-D) because its petrophysical properties do not change in any spatial direction (dimension) around the logging tool.

Earth models used in describing thinly bedded reservoirs can be categorized in terms of their *dimensionality*, ranging from the simplest (0-D) to the most complex (3-D, or three-dimensional). They can also be categorized as either *explicit* or *statistical*. In an *explicit* earth model, each earth region consists of a single petrophysical bed type. In a *statistical* earth model, each region comprises two or more bed types whose relative frequency and geometric arrangement are described in statistical terms.

1-D earth models

In an explicit one-dimensional (1-D) earth model, petrophysical properties vary only along a single dimension. See Figure 10.1.

- In an infinitely thick homogeneous formation, borehole and invasion effects can be modeled as 1-D variations along the radial dimension perpendicular to the borehole (case A in Figure 10.1). Petrophysical properties needed for tool-response modeling can be expressed as a function of the radial variable r in cylindrical coordinates; i.e., $P = P(r)$.
- A layered geometry of planar, homogeneous sedimentary beds implies that the petrophysical properties of the formation vary only in the z-coordinate direction perpendicular to bedding. If variations around the borehole are excluded, this geometry forms a 1-D model (case B in Figure 10.1). Petrophysical properties needed for tool-response modeling can be expressed as a function of z alone; i.e., $P = P(z)$
- Case C in Figure 10.1 represents an explicit earth model that is not strictly one-dimensional. However, most 1-D numeric simulation models can be adapted to handle both dipping planar beds, and plane boundaries between regions with different dip angles.

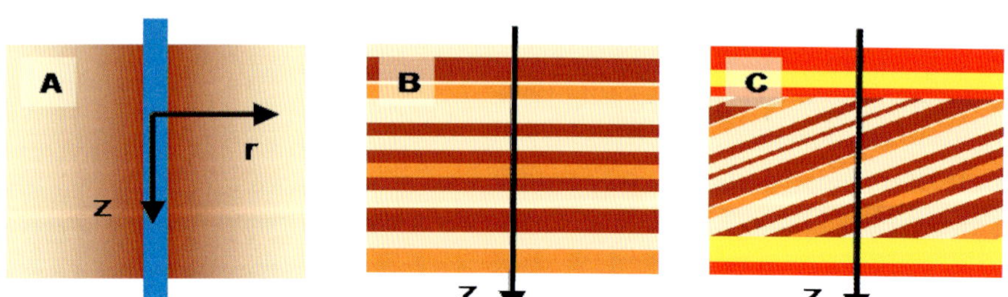

Figure 10.1. One-dimensional (1-D) earth models. A: Variation in radial (r) direction only. B: Variation in borehole (z) direction only. C: Not strictly 1-D, but most numeric tool models can treat as 1-D.

2-D earth models

If the logging environment varies in the vertical and radial directions due to bedding, borehole, and mud filtrate invasion but is reasonably axisymmetric around the borehole, then the earth model can be adequately defined as two-dimensional (2-D). Petrophysical properties needed for tool-response modeling can be expressed as a function of radius (r) and borehole dimension (z) in cylindrical coordinates; i.e., P = P(r, z). See Figure 10.2.

In general, numeric simulation methods are required to model log responses in explicit two-dimensional earth models.

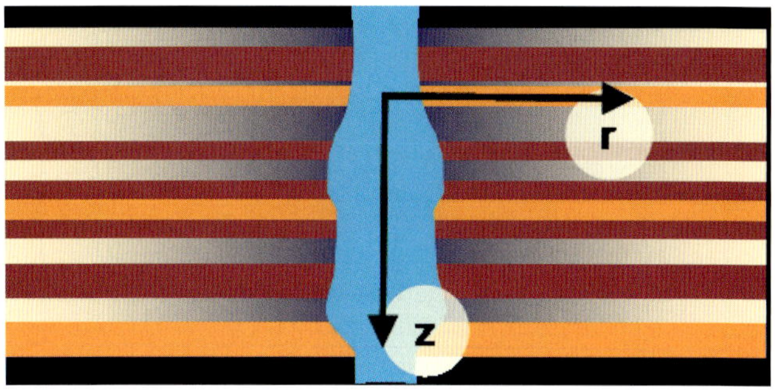

Figure 10.2. Two-dimensional (2-D) earth model.

3-D earth models

If three independent variables are required to describe the variation in petrophysical properties within the region of interest, then the earth model is three-dimensional (3-D), i.e., P = P(x, y, z). Figure 10.3 shows a combination of dipping beds, filtrate invasion, and borehole effects that requires a 3-D earth model. In the figure the x-axis is along the direction of maximum dip and the beds are sliced along the x-z plane.

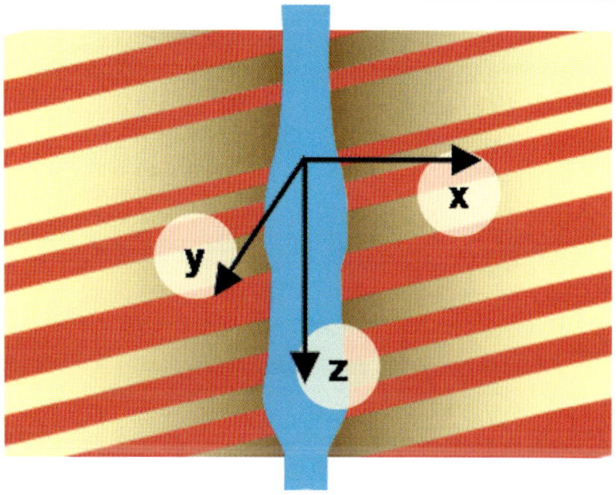

Figure 10.3. Three-dimensional earth model.

Numeric simulation methods are required to model log responses in fully three-dimensional earth models. These 3-D simulations require a level of computing power and time that has (so far) prohibited their practical application in thin-bed analysis.

Statistical earth models

Figure 10.4 illustrates the difference between an *explicit* 1-D earth model and a *statistical* earth model representing the same thinly bedded reservoir interval. In the explicit model, each individual thin bed is described by its unique top depth, base depth, and petrophysical properties. In the statistical model, the thinly bedded interval is described as a single unit with top and base depth. The unit comprises two petrophysical bed types, each described by its volume fraction and its average petrophysical properties. The statistical model contains no information about the location or thicknesses of individual beds within each petrophysical bed type.

A statistical model representing a more complex geometry is shown in Figure 10.5. The explicit model on the left is highly complex and three-dimensional, comprising a set of aligned oblate ellipsoids of revolution which approximate beds of finite lateral extent. In contrast, the statistical earth model on the right is essentially zero-dimensional, even though some of its petrophysical properties are anisotropic. Only the volume fraction and the aspect ratio of each petrophysical bed type are required to capture the bedding geometry in statistical terms. This statistical model becomes 1-D if bounding beds are added above and below the interval pictured.

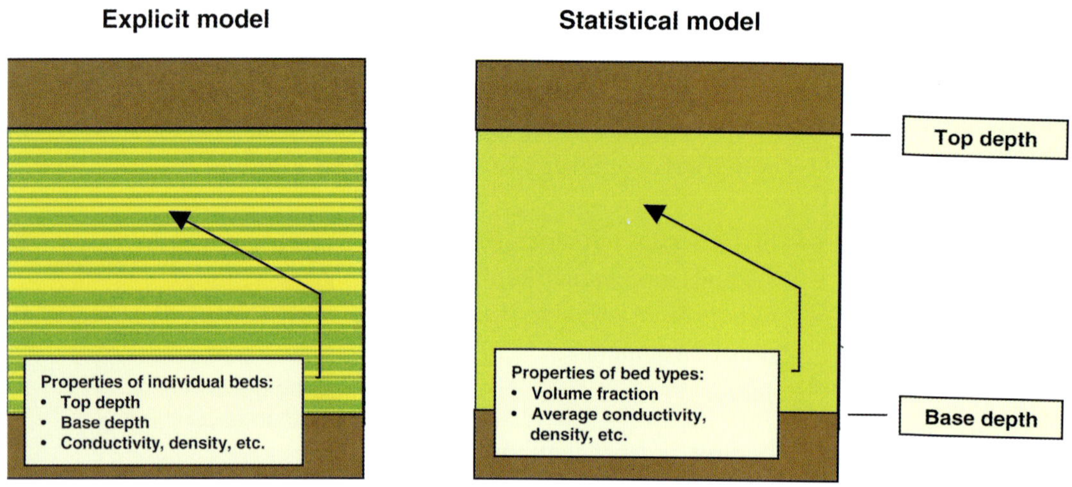

Figure 10.4. Comparison of explicit and statistical earth models for a thinly bedded reservoir interval with parallel planar beds.

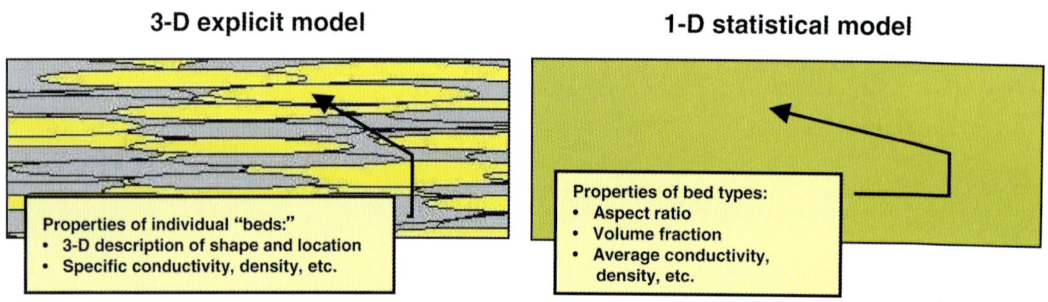

Figure 10.5. Comparison of explicit and statistical earth models for a thinly bedded reservoir with beds of finite lateral extent.

Forward and Inverse Modeling: Introduction

Forward modeling

The process of calculating a well-log response numerically for a given earth model is referred to as *forward modeling*. An algorithm or computer code for performing such a calculation is a *forward model*. Each forward model is closely linked to a specific type of earth model, since it is the earth model that carries all the data required as input for the forward model's calculations. Forward models typically produce a unique solution at each depth of the earth model, so they may also be called *tool response functions*.

Figure 10.6 is a schematic illustration of 1-D forward models of shallow, medium, and deep resistivity logs. The geometry of the earth model is shown on the left. The resistivity of the earth model's beds is shown on the right as the orange line and shading, labeled R_T. The forward models assume no borehole or invasion effects. These forward models, which are 1-D convolution filters, are shown graphically in the middle. Overlaid on the R_T curve on the right-hand side are the shallow (blue), medium (green), and deep (red) resistivity logs calculated by the respective forward models.

Several types of forward model useful in thin-bed evaluation are described below.

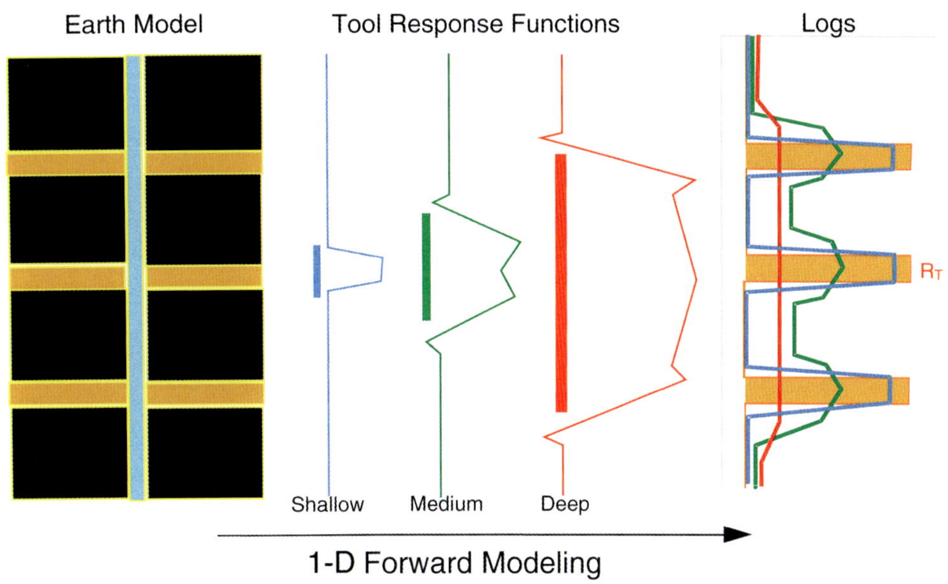

Figure 10.6. Schematic representation of 1-D forward modeling.

Inverse modeling

When we log a well, what we hope to obtain from the well logs is an accurate *earth model* of the logged formations, including both the bedding geometry and the petrophysical properties of the beds. The logs themselves are only indirect indicators of this true earth model because their measurements are affected by bed thickness, shoulder beds, borehole conditions, invasion, and other factors. Therefore the estimation of the earth model from the logs requires a process of inference. This process of deriving the geometric and petrophysical parameters of an earth model from a set of given field logs and the appropriate tool-response functions is referred to as *inverse modeling* or simply *inversion.*

Figure 10.7 illustrates inversion of a deep resistivity log using an iterative procedure based on the 1-D tool response function shown in Figure 10.6.

In Figure 10.7, the earth-model geometry (horizontal planar beds) is shown on the left. In the center, the log at A shows an initial guess for the true resisistivities of the earth model as the squared orange curve. The measured field log is shown in blue, and the calculated log based on the earth model and the tool response function is shown in red. Mismatches between the field log and the modeled log are evident.

In the inversion step, the resistivity values of the earth model's beds are adjusted iteratively to optimize the match between the field log and the modeled log. The log at B shows the final inversion result. The revised earth-model resistivities are again shown as the squared orange curve. The calculated log (red) hides the field log (blue), showing that a good replication of the measured data was obtained. The final earth model (with its beds and resistivities) is then taken to be a better representation of the actual distribution of R_t — true formation resistivity — than the original measured log.

In the example of Figure 10.7, the iterative inversion procedure adjusted the earth-model resistivities and left the bed boundaries fixed. This procedure might be employed in a case where bed boundaries can be defined with confidence using high-resolution data such as a borehole image log [Chapter 9]. In general, however, all parameters of the earth model, including the number of beds and the locations of their boundaries, can be adjusted to optimize the fit between the modeled and measured logs.

Several methods of inverse modeling that have been employed in thin-bed evaluation are discussed below.

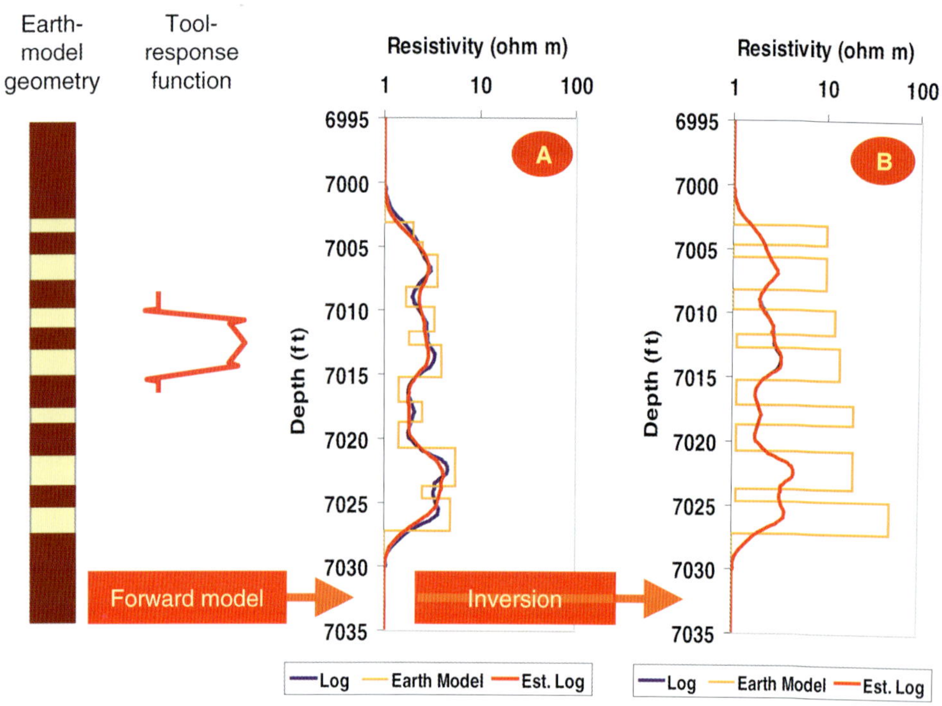

Figure 10.7. Schematic illustration of inverse modeling [modified from Passey et al., 2004].

Types of Forward Models

Overview

Forward models for logging tool responses are the foundation of any method for thin-bed evaluation. Our integrated approach employs a variety of models that differ widely in their level of detail and their computational requirements. These models may be classified into four categories, listed here roughly in order of decreasing complexity.

1) Numeric simulations of tool physics in one, two, or three dimensions;

2) Simplified approximate solutions to tool physical equations;

3) One-dimensional convolution models;

4) Interval-average models.

These categories of forward model are introduced in general terms below.

Numeric simulations

For our discussion of thin-bed analysis we consider detailed numeric simulations only for resistivity logging tools (i.e., not for porosity or lithology tools). Resistivity tool response modeling belongs to the class of so-called boundary-value problems. A tool with its transmitter(s) and receiver(s) located at specified points is introduced into an earth model. The transmitters excite the earth model medium in some specified manner. Maxwell's equations describe the resulting electromagnetic field at every point in space. The earth model specifies the boundaries of regions of differing resistivity and the resistivity within each region. Solving Maxwell's equations for points within the tool's support volume is then a boundary-value problem.

With enough simplifying assumptions about the spatial distribution of resistivity, the solutions of Maxwell's equations for the boundary-value problem may take the form of simple formulas approximating the electromagnetic field for any desired point in space. Such formulas belong to the second category listed above.

For more complicated geometries the electromagnetic fields cannot be expressed in terms of analytical functions, but must be computed numerically at many points in space simultaneously. The required computing power increases significantly as the dimensionality of the earth model increases from one to three. For examples of 1-D and 2-D numeric simulations of resistivity log response, see Anderson and Gianzero [1983], Anderson [1986], Anderson and Barber [1988], Anderson et al. [1995], Chemali et al. [1983; 1987], Gianzero et al. [1985], and Kennedy [1995].

To meet the needs of thin-bed analysis, numerical simulations must be sufficiently fast and efficient so that it is practical to incorporate the forward model into an inversion procedure as described above. This requirement excludes full 3-D models from further consideration here. Only a few 3-D modeling studies have been published; see, for example, Anderson et al. [1996], Graciet and Shen [1997], Wang et al. [1998], Anderson [1988], and Yin et al. [2002].

Approximate solutions

The category of approximate solutions refers to any forward model that approximates a log response as a relatively simple, closed-form function of earth model parameters. An example is Equation 6.8, which describes induction log response in a homogeneous transversely isotropic medium as a function of the parallel and transverse conductivities of the medium and the dip angle of the borehole relative to the medium.

1-D convolution models

One-dimensional convolution functions are a particular kind of approximate solution that is defined for earth models like those illustrated in Figure 10.1(B), whose petrophysical properties vary only in the direction parallel to the borehole.

Figure 10.8 illustrates a 1-D convolution model for a gamma ray (GR) log. The earth model (a 10-ft [3-m] subinterval of a longer interval) is subdivided into layers of equal thickness. The earth model is shown on the left and its subdivision into 10 layers per ft is indicated by the tick marks on the left vertical axis. Petrophysical beds, indicated by colored bars, are composed of one or more contiguous layers. The orange beds have GR = 20 and the green beds have GR = 100, as indicated on the horizontal axis. The set of GR values situated at the center of each layer form a data series, $\{GR_i, 1 \le i \le 100\}$.

The convolution model is defined by a *convolution filter* representing the logging tool's *impulse response function* — that is, its vertical response to an infinitesimally thin bed. The convolution filter quantifies how the finite-length logging tool "smears" the measurement of a very thin bed. Mathematically the convolution filter is expressed as a set of constant numeric coefficients, $\{f_i, -N \leq i \leq N\}$, which are normalized so their sum is one (1). The index (i) represents the same depth increment as the layer thickness in the subdivided earth model. In Figure 10.8 the convolution filter is plotted in the center with its indices along the vertical axis. Its non-zero coefficients extend from about i = –20 to i = 20.

The modeled log, GR(mod), is computed by *convolving* the earth-model data series with the tool filter coefficients as in Equation 10.1.

$$GR(mod)_k = \sum_{i=-N}^{N} f_i \cdot GR_{k-i} \tag{10.1}$$

For this example, depth is related to the index (k) in Equation 10.1 by

$$\text{depth} = XX00 + (k - 1)/10 \qquad \text{where } 1 \leq k \leq 100 \tag{10.2}$$

The blue dot on the modeled GR log (right side of Figure 10.8) represents $GR(mod)_{51}$, located at depth XX05 feet. The other modeled points are plotted as a continuous blue curve.

Convolution models can be developed for all the most commonly-used conventional logs [Looyestijn, 1982]. In general, logging tools are designed so that their response characterisitics are as close as possible to *linear* for the simple earth geometry illustrated in Figure 10.1(B). Linearity in this context means precisely that the tool's vertical response can be approximated adequately by a 1-D convolution model.

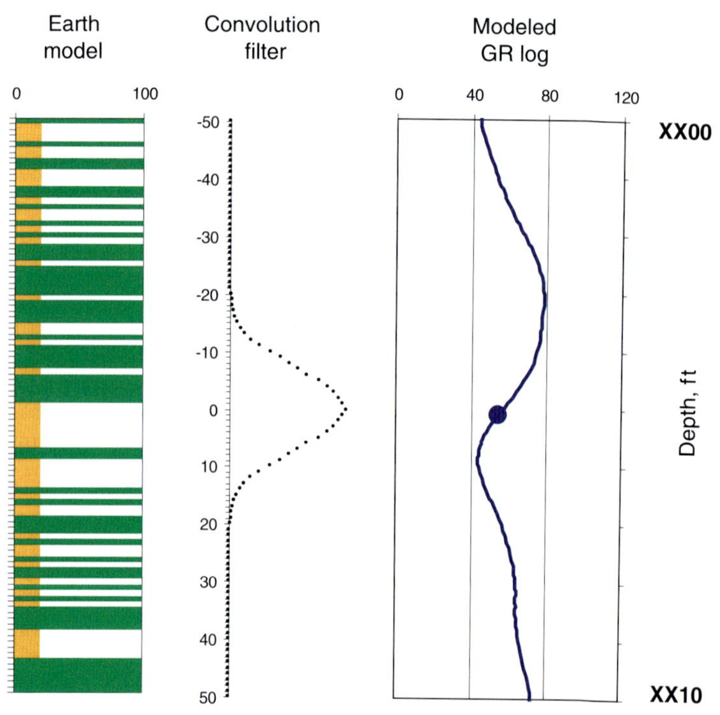

Figure 10.8. The components of a 1-D convolution model for a GR log.

Interval-average models

The tool response functions discussed above all calculate a log value representing a single measurement at a given depth. We are also interested in a coarser kind of model that calculates the average of log measurements across a defined depth interval. These interval-average models are valid under the same set of conditions as the 1-D convolution models: that is, conditions where the tool response is sufficiently linear and where borehole and invasion effects are not significant (or can be corrected in advance).

The components of a simple interval-average model for a bulk density log are depicted in Figure 10.9. For this illustration we consider the interval from 9010 to 9030 ft [2746 to 2752 m]. The earth model, shown on the left, has two bed types. The relevant petrophysical characteristics of the earth model and their values over the defined interval are summarized in the table on the right of the figure.

The interval-average model for the bulk density over the defined interval is given by Equation 10.3, where D_{int} is the average of the bulk density log values over all the logged depths in the interval. Note that Equation 10.3 uses only the volumetric distribution of bed types and their values; it does not use the detailed bed geometry. The earth model as depicted in Figure 10.9 shows more detail than the interval-average model actually requires.

$$D_{int} = v_1\rho_1 + v_2\rho_2 \qquad (10.3)$$

Despite its simplicity, Equation 10.3 is a non-trivial approximation. The right side of Equation 10.3 is the actual average bulk density of the earth model across the defined interval; its value for the example is 2.281 g/cm^3. The left side is the average of several log measurements whose support volumes overlap; its value for the example is 2.284 g/cm^3. Thus the model error for the example is 0.003 g/cm^3. If we assume that the log's vertical response is accurately represented by a 1-D convolution filter, then it can be demonstrated that the error in Equation 10.3 is introduced at the top

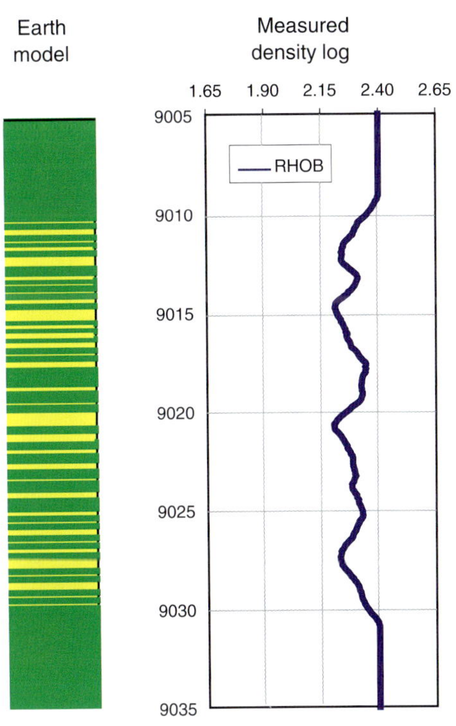

Earth model characteristics		
Bed Type	Volume fraction	Bulk density
1 (Yellow)	$v_1 = 0.49$	$\rho_1 = 2.155$
2 (Green)	$v_2 = 0.51$	$\rho_2 = 2.403$

Figure 10.9. Components of an interval-average model for bulk density. Depth in ft.

and base of the defined interval, where a few of the measurements are influenced by the formation outside the interval. In general the magnitude of the error can be approximated by

$$e \approx \frac{1}{2} \frac{h_{tool}}{h_{int}} |\Delta L| \tag{10.4}$$

where h_{tool} is the length of the tool's impulse response function, h_{int} is the length of the defined interval, and ΔL is the difference between the average log values inside and outside the defined interval. For the example h_{tool} is 1.8 ft [.5 m], h_{int} is 20 ft [6 m], ΔL is 0.119 g/cm³, and the estimated error is 0.005 g/cm³.

Note that Equation 10.3 does not depend on the detailed geometry of the earth model, i.e., the locations of bed boundaries. In effect the interval-average model replaces an *explicit* earth model with a *statistical* one involving only bed-type distributions, average bed-type petrophysical properties, and averaged log values. This makes the interval-average model very useful for analyzing reservoirs whose beds are too thin to be defined even by high-resolution log data [Chapter 12].

Methods of Inversion

Deconvolution

The term "deconvolution" has sometimes been used loosely in the well-logging literature to refer to any kind of inverse modeling, but the term has a strict meaning in signal-processing theory. To describe this meaning we first need to introduce standard symbolic notation for the convolution operation illustrated in Equation 10.1. Let $D(n) = \{d_n, -\infty \leq n \leq \infty\}$ be a data series representing the values of a petrophysical variable in an earth model with uniform layers, like $\{GR_n\}$ in Equation 10.1. Let $G(n) = \{g_n, -\infty \leq n \leq \infty\}$ be the convolution filter series representing a logging tool's geometric response function, like that shown in Figure 10.8, with the coefficients normalized so that $\Sigma_n g_n = 1$. Finally let $L(n) = \{l_n, -\infty \leq n \leq \infty\}$ be the data series representing the modeled log, like GR(mod) in Equation 10.1. Then the convolution of D(n) with G(n) to produce L(n) is denoted with an asterisk:

$$L(n) = G(n) * D(n) = \sum_{i=-\infty}^{i=\infty} g(i) \cdot d(n-i) \tag{10.5}$$

Now, deconvolution means to find and apply a new filter series H(n) that is (or approximates) the inverse filter for G(n). In other words, H(n) has the property that

$$D(n) = H(n) * L(n) \tag{10.6}$$

for every pair of data series, D(n) and L(n), that satisfy Equation 10.5. Equation 10.6 indicates that the petrophysical values of the earth model, D(n), may be recovered from the measured log, L(n), by simply applying a linear filter, namely the inverse filter H(n). And this inversion is accomplished without knowing the distribution of petrophysical beds!

For a period during the 1970s and 1980s, deconvolution seemed to be the holy grail of log resolution enhancement and was the subject of many publications. Most of these focused on induction tools [e.g., Thadani and Merchant, 1982; Howell and Fisher, 1982; Woodhouse et al., 1984] but others explored the application of deconvolution to other common logs [e.g., Czubek, 1973; Branisa, 1974; Morland, 1984].

Ultimately it was found that deconvolution filters have limited practical utility because convolution models for well logs are approximations, and the derived deconvolution filters tend to introduce excess noise and artifacts in the inverted data series D(n) [Looyestijn, 1982; Barber, 1988]. Nonetheless, *convolution models* remain a valuable forward-modeling technique that can be incorporated in other inversion schemes, such as those discussed below.

Iterative inversion

Iterative inversion is the main inversion method used in well-log modeling. The generic process of iterative inversion is illustrated as a flowchart in Figure 10.10. The forward model is computed for an initial earth model and the modeled log is compared to the measured log. If the agreement is not "close enough," the parameters of the earth model are changed to move the modeled log closer to the real data. This process is iterated until the agreement is "close enough." The final result is a "squared" inverted log which consists of the bed boundaries and log values of the final earth model, like the rectilinear orange log shown in Figure 10.7(B).

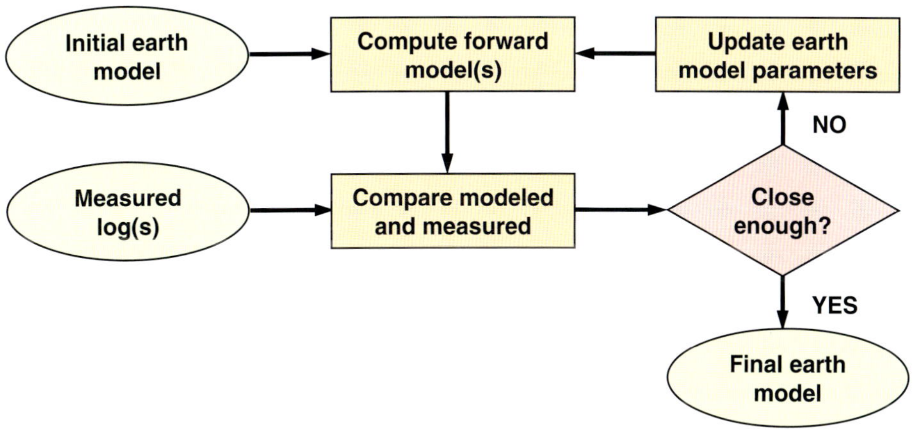

Figure 10.10. Iterative inversion of a well-log forward model.

This general inversion scheme has many variants, which can mostly be defined through combinations of the following five categories.

1) Earth model. The earth model may be 1-D, 2-D, or 3-D. As the dimension of the earth model increases, the number of earth-model parameters and the complexity of the forward model also tend to increase, leading to longer computing times for inversion.

2) Number of logs inverted. A single log may be inverted, or several may be inverted simultaneously. For example, it is common to invert shallow, medium, and deep-reading resistivity logs simultaneously to resolve a 2-D earth model where borehole, invasion, and bedding effects are all present [Chapter 11].

3) Manual or automatic updating. Earth model parameters may be modified manually by the analyst until a satisfactory visual or quantitative fit is obtained. Alternatively the process of updating may be automated and a quantitative criterion may be established to define the "good enough" fit between measured and modeled logs.

4) External constraints. In the inversion process all the parameters of the earth model may be subject to change, including the geometric parameters and the petrophysical parameters. Alternatively some of these parameters may be kept fixed as a way of incorporating external information into the inversion. One example is the incorporation of fixed bed boundaries obtained from a high-resolution log. Another example would be the imposition of a zero invasion depth in the shale beds of a 2-D resistivity model.

5) Type of forward model. The forward model(s) in the inversion may be of the numeric, analytic, convolution, or interval-average type(s). For a fully automated inversion procedure it is important to optimize the computing speed of the forward model. A very complex numeric simulation, such as would be required for a three-dimensional forward model, may not lend itself well to an automated inversion procedure.

Specific applications of this general scheme will be presented in Chapter 11.

Monte Carlo inversion

So far, the discussion of inversion has implicitly assumed that we expect to find a single unique solution when we set out to invert a set of well logs. However, as has already been pointed out and will be reiterated in subsequent examples, the presence of beds thinner than well-log resolution implies there is no unique solution to our inversion problem. Therefore the presence of thin beds introduces a significant level of *uncertainty* into the evaluation of reservoir hydrocarbon pore-thickess.

The problem of inverting a set of well logs for a given earth model may be classified as a *discrete inverse problem*. It is "discrete" because the earth model is defined by a finite number of discrete parameters, as opposed to an infinite continuum of parameters. For a discrete inverse problem it is possible to consider applying the technique of *Monte Carlo inversion* [Tarantola, 1987], which is a special type of iterative inversion that explicitly addresses the issue of uncertainty.

Three categories of uncertainty can be incorporated into the analysis that leads to Monte Carlo inversion:

1) Uncertainty in the parameters of the geologic model;
2) Uncertainty in the measured log data;
3) Uncertainty in the forward model(s) — i.e., uncertainty due to the approximate nature of the forward model.

Figure 10.11 illustrates these uncertainties using a simple hypothetical forward model. The graph on the left shows on its left vertical axis a histogram representing the measurement and its uncertainty, and on its bottom horizontal axis a histogram representing the initial distribution of the model parameter (labeled "System parameter"). These initial or *a priori* distributions depict the model parameter and the measurement as being independent and unrelated. The cluster of blue dots in the center of the graph represents their independent joint distribution.

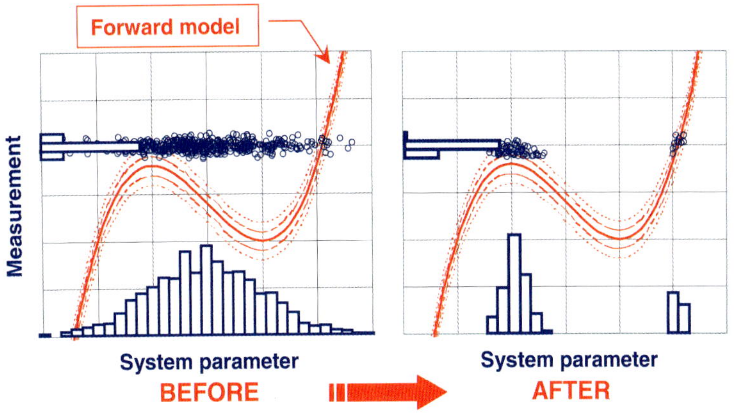

Figure 10.11. Schematic representation of measurement and model uncertainties before and after imposition of the constraint imposed by the forward model. Modified after Tarantola [1987], his figure 1.10.

The red curve represents the forward model, a function that calculates the measurement as a function of the model parameter and thereby imposes a constraint on their joint distribution. The dotted envelope around the red curve represents the uncertainty in the forward model.

The forward model imposes a constraint on the joint distribution of the measurement and the model parameter. In particular, referring to the joint distribution in the left graph of Figure 10.11, the points that don't fall within the uncertainty envelope of the forward model are not consistent with the forward model. The method of Monte Carlo inversion provides a way to impose the forward-model constraint and to derive new distributions that are consistent with both the forward model and the *a priori* distributions. The new distributions resulting from the application of Monte Carlo inversion are shown in the graph on the right of Figure 10.11. Note how the possible values for the model parameter are now significantly constrained.

Figure 10.12 outlines the method of Monte Carlo inversion in a flowchart that illlustrates its connection to the general scheme of iterative inversion (Figure 10.10). The differences are outlined below.

The first difference is that the initial earth model is defined probabilistically. This means that an *a priori* distribution, rather than a single average value, must be specified for each model parameter.

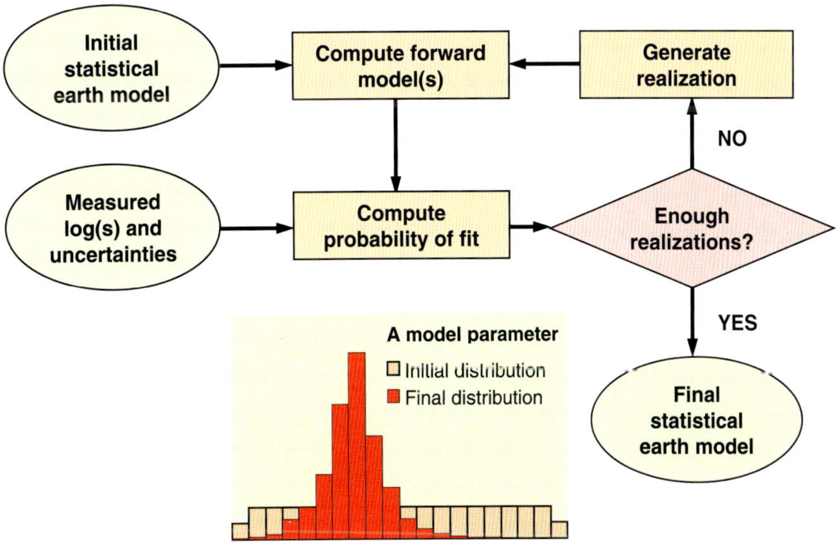

Figure 10.12. Monte Carlo inversion of a well-log forward model.

The *a priori* distribution reflects everything that is known about the parameter before taking into account the constraints imposed by the logs and their forward models. As a concrete example, we might know that reservoir sand porosity is between 0.18 and 0.30, without any preferred value in that range. This knowledge translates as a uniform porosity distribution between the bounds of 0.18 and 0.30. The tan histogram in the graph at the bottom of Figure 10.12 illustrates a uniform a priori parameter distribution.

Second, the forward model is calculated on a *realization* of the statistical earth model. A realization is simply a set of model parameter values that are drawn at random from the *a priori* distributions of the initial model. Each new realization is independent from the previous one; there is no attempt to move in a direction of improved fit between modeled and measured logs.

Third, instead of testing to see if the fit between modeled and measured logs is "good enough," we compute a *probability* of fit. Since there is uncertainty in both the measured log(s) and the forward model(s), any computed model value has some finite probability of being statistically indistinguishable from the measured value. This probability is then used to weight the current realization in building a final set of constrained parameter distributions.

Fourth, the test for completion of the Monte Carlo inversion is simply whether enough realizations have been generated to clearly define the final constrained parameter distributions. Since many of the realizations may receive near-zero weight because of their low probability of fit, it may take tens of thousands of realizations to define the constrained distributions.

Finally, the result of the Monte Carlo inversion is a new set of distributions for the earth-model parameters, which have now been subjected to the constraints imposed by the log measurements and their forward models. The two histograms graphed at the bottom of Figure 10.12 illustrate how a model parameter (for example, sand porosity) can go from an initial, poorly constrained, uniform distribution to a much narrower and more peaked final distribution. These final distributions integrate all the available information about the model parameters: The *a priori* knowledge, the constraint imposed by the measured logs and their forward models, and the uncertainties in the measurements and models. Most importantly, the final distributions describe explicitly both the *solution* for the model parameters and the *uncertainty in the solution.*

The number of realizations required to adequately sample the joint distribution of model parameters increases exponentially with the number of parameters. Interval-average forward models have a manageable number of parameters and lend themselves to Monte Carlo inversion. This is the basis for the method we call Volumetric Laminated Sand Analysis [Chapter 12].

Models for Lithology and Porosity Logs

Convolution models

The convolution filter for a given logging tool, also called its *geological impulse response function*, depends on the physics of the tool's measurement and on the specific geometry of its design, e.g., the spacings between sources/transmitters and detectors/receivers. Thus, the appropriate filter for a specific version of a specific service company's tool of a given type depends on factors unique to that tool's design. It is beyond the scope of this publication to present a catalog of tool-specific filters or a comprehensive discussion of methods of designing such filters. Here below are some general considerations, followed by tool-specific references and examples.

1) Non-linearity in a logging tool's response can be viewed as variability in its impulse function — that is, the coefficients of the convolution filter vary depending on the geology of the logged formation. Thus depending on the degree of nonlinearity, a filter designed for an "average" set of geologic conditions may provide an approximation that is useful for forward modeling and iterative inversion (but not for strict deconvolution) across a wider range of conditions.

2) To model a real field log, a complete convolution filter must incorporate the tool's inherent geologic impulse response and must also duplicate any post-acquisition filtering. Nuclear and NMR logs, in particular, are smoothed digitally to reduce the effect of detection statistics, and this smoothing must be incorporated in the convolution filter.

3) The convolution filter and the earth model must be sampled at the same depth increment, which may differ from the tool's standard sampling increment. Therefore in practice it may be necessary to resample a given response function at a variety of depth increments.

Gamma-ray tools. Convolution modeling works well for gamma-ray logs. See Czubek [1973; 1986] and Morland [1984] for thorough expositions on the derivation of impulse-response functions for total and spectral gamma-ray tools. Convolution filters for a logging service company's specific gamma-ray tool may be available from the service company. A theoretically-derived impulse response function for a 1-ft [.3-m] GR detector is illustrated in Figure 10.13. This function does not include the effect of post-acquisition filtering.

The precise impulse response for a GR tool depends on formation bulk density and on borehole fluid density. In practice, since these densities vary, the GR convolution model is always approximate.

Spontaneous potential. The spontaneous potential log has a complex and highly non-linear response that is not amenable to simple convolution modeling. See Glowinski et al. [1988] for a description of a detailed numerical simulation of the SP and its application to inversion and interpretation. SP inversion requires formation resistivity as an input.

Acoustic logs. The basic impulse response of an acoustic tool is a rectangular "boxcar" function, indicating that the measured transit time is a simple average of the transit times of the layers between the transmitter and receiver (or between the two receivers for a borehole-compensated tool) [Looyestijn, 1982]. An impulse response for an acoustic tool with 2-ft [.6-m] spacing is illustrated in Figure 10.13.

Density and neutron logs. Density and neutron porosity tools employ dual detectors at differing source-to-detector spacings to provide compensation for borehole and mudcake effects. Countrates from these detectors are combined via non-linear algorithms to product the tools' characteristic output. The use of convolution filters to model these tools, particularly the older versions, has been problematic [Looyestijn, 1982]. However, in the late 1980s the logging service companies began to address the need to optimize the vertical response characteristics of these tools [Galford et al., 1986; Flaum et al., 1989; Jacobson et al., 1990]. The result was improved processing methods that enhanced the resolution of the tools' basic measurements, and also improved their vertical symmetry and linearity so that their responses might be better approximated by convolution filters. Convolution models for specific tools may be obtained from the service companies.

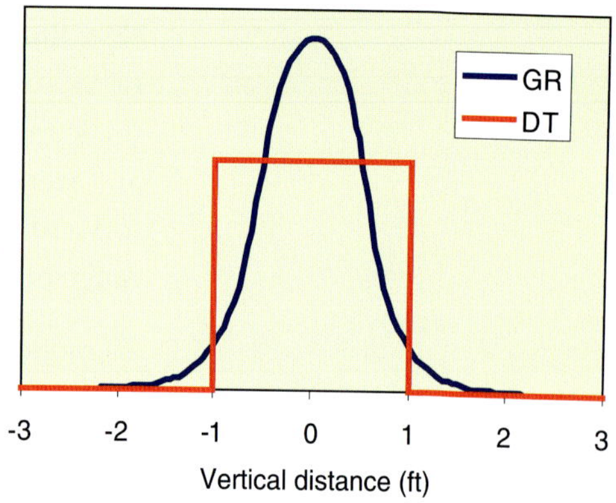

Figure 10.13. Impulse response functions for a 1-ft [.3-m] gamma ray detector (GR) and an acoustic tool with 2-ft [.6-m] spacing (DT).

Interval-average models

Interval-average models for logs measuring isotropic properties (e.g., gamma ray, bulk density, neutron porosity) all resemble the example described previously for bulk density. To describe these models in general terms, let L be the measured log and let L_{int} be its average across all the measure points in the defined interval. Let $v_1, \ldots, v_N$ be the volume fractions of the N bed-types within the defined interval, and let $L_1, \ldots, L_N$ be the earth-model values of the given log for each of the bed-types. Then the volume-weighted interval-average model for the log is defined by Equation 10.7.

$$L_{int} = \sum_{i=1}^{N} v_i L_i \tag{10.7}$$

Some log measurements, such as the photoelectric factor and natural gamma ray, are properly averaged using mass fractions instead of volume fractions as weights. In particular, Equation 10.7 should be applied to the volumetric photoelectric factor (U) and not directly to the photoelectric factor log (Pe) [Gardner and Dumanoir, 1980]. Even though gamma-ray detector response properly depends on mass fractions, Katahara [1995] has shown that Equation 10.7 is usually an adequate approximation for the gamma-ray log.

The error in Equation 10.7 due to the effects of shoulder beds at the top and base of the defined interval is approximated as in Equation 10.8.

$$\text{error} \approx \frac{1}{2}\frac{h_{tool}}{h_{int}}|\Delta L| \tag{10.8}$$

In Equation 10.8, h_{tool} is the length of the tool's impulse response function, h_{int} is the length of the defined interval, and ΔL is the difference between the average earth-model log values in the defined interval and in the shoulder beds. If ΔL is large then the error may be reduced by computing the log average, L_{int}, over a subinterval of the defined interval, omitting one-half the tool length (h_{tool}) at the top and base to avoid the effects of the shoulder beds.

Models for Resistivity Logs

Introduction

Deep-reading resistivity logs (induction logs and focused-current contact logs, such as laterologs) have the poorest vertical resolution of all the conventional logs utilized in thin-bed evaluation. Therefore the modeling of resistivity tools is a critical component of our approach, and we use models of all levels of detail and complexity.

Numerical simulation models for resistivity logs were introduced previously, and the practice of forward modeling and iterative inversion using numerical models is discussed in some detail in Chapter 11. The objective of this section is to introduce several *approximate, 1-D convolution*, and *interval-average* models for resistivity tools.

Approximate solutions

Borehole perpendicular to bedding. As discussed in Chapter 6, the resistivity (or conductivity) in a thinly bedded reservoir exhibits *macroscopic anisotropy*. We model the conductivity in these reservoirs as *transversely isotropic*, meaning that it can be described by two components: σ_p, the conductivity parallel to bedding planes, and σ_t, the conductivity transverse (perpendicular) to bedding planes. Figure 10.14 illustrates the situation where a borehole penetrates perpendicular to uniformly-spaced thin beds.

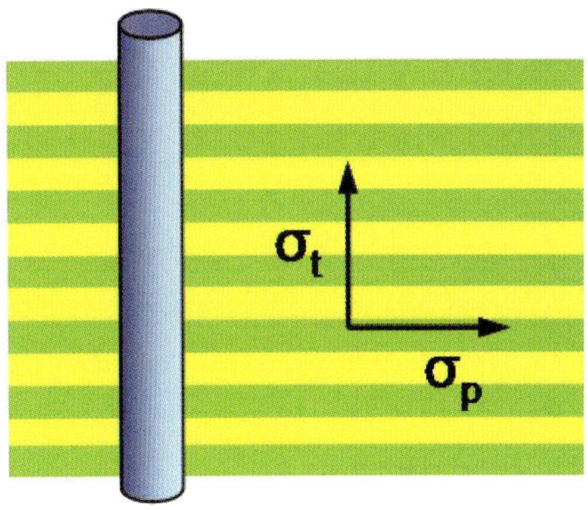

Figure 10.14. Borehole perpendicular to uniformly spaced thin beds.

The average parallel and transverse conductivities across a thin-bedded interval are expressed in terms of the volume fractions of the bed-types ($v_1,\ldots,v_N$) and the conductivities of the bed-types ($\sigma_1,\ldots, \sigma_N$) in Equations 10.9 and 10.10, which generalize Equations 5.5 and 5.6.

$$\sigma_p = \sum_i v_i \sigma_i \qquad \text{(parallel conductivity)} \qquad (10.9)$$

$$\sigma_t = \left(\sum_i v_i \ \sigma_i^{-1} \right)^{-1} \qquad \text{(transverse conductivity)} \qquad (10.10)$$

Deep induction: *Borehole perpendicular to bedding.* Anderson [1986] showed using numeric simulations that, when beds are 1-ft [.3-m] thick or less in the situation of Figure 10.14, the induction tool measures the parallel conductivity, σ_p. Thus we have the approximate solution for deep induction-log resistivity (R_{ILD}) given by Equation 10.11.

$$R_{ILD} = \frac{1}{\sigma_p} \qquad (10.11)$$

Deep laterolog: *Borehole perpendicular to bedding.* The same approximation holds for the deep laterolog when the beds are sufficiently thin. Figure 10.15 illustrates a benchmark numerical simulation of the dual laterolog (DLL) tool's response for a series of 3-in. [8-cm] beds.

The beds in Figure 10.15 alternate in resistivity between 1 and 20 ohm m. The effective parallel conductivity is 0.525 mmho/m, corresponding to 1.9 ohm m resistivity. Figure 10.15 shows that in the center of the thinly bedded interval (away from shoulder effects), the deep laterolog (LLD)

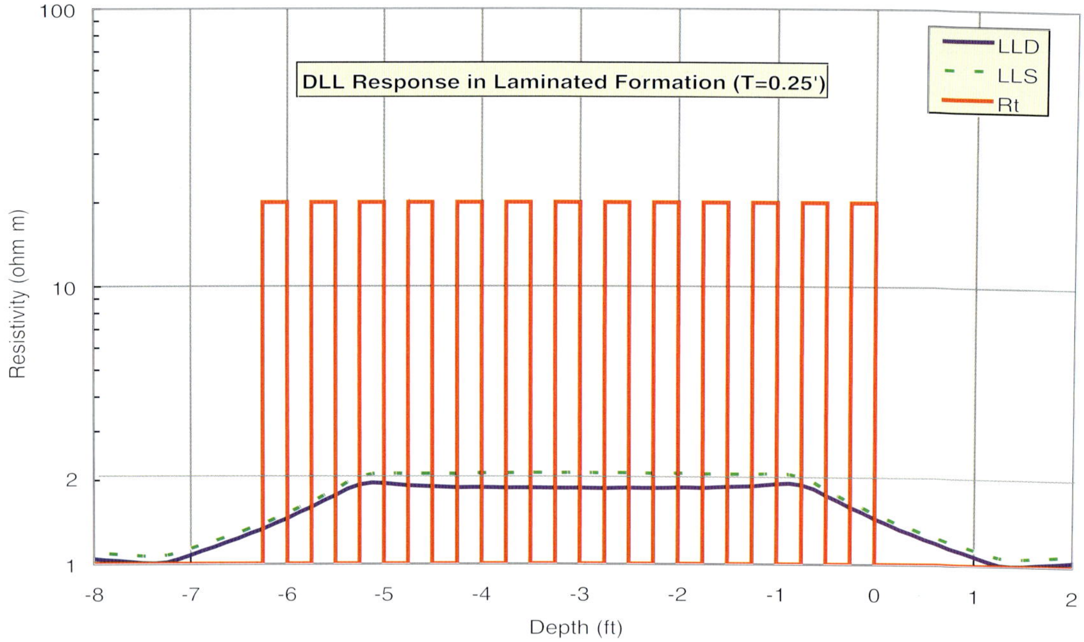

Figure 10.15. Simulation of dual laterolog in thin-bedded formation.

measures this parallel average. Thus the approximate solution for LLD response perpendicular to thin beds is given by Equation 10.12.

$$R_{LLD} = \frac{1}{\sigma_p} \qquad (10.12)$$

The shallow laterolog (LLS) resistivity in Figure 10.15 is elevated slightly from the parallel average because the shorter LLS current path cuts across bedding and is more influenced by the transverse conductivity, σ_t.

A benchmark DLL simulation with 6-in. [15-cm] beds produces similar results but the curves are bumpier, reflecting the fact that the dual laterolog has finer vertical resolution than the deep induction log.

Induction: *Dip effects.* Figure 10.16 illustrates the situation where a borehole penetrates a thin-bedded formation with relative dip angle θ. Induction logging tools measure conductivity via induced current loops concentric with the borehole. Therefore, the direction of current flow is perpendicular to the borehole, and the orientation of the current loops relative to bedding planes depends on the relative dip angle θ. Figure 10.16 shows the apparent conductivity measured by the induction tool as σ_{IL}.

Figure 10.16. Borehole penetrates thin-bedded formation with dip angle θ. σ_{IL} = induction-log conductivity.

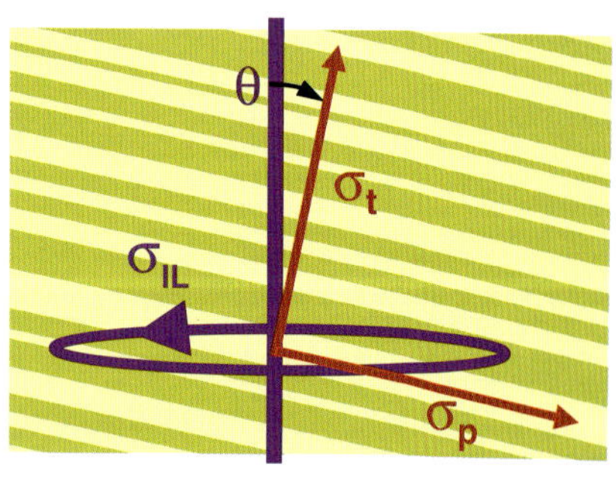

Moran and Gianzero [1979] determined the theoretical response of the induction tool in a dipping anisotropic formation. Equation 10.13 is the analytic expression for this ideal response in terms of θ, σ_p, and σ_t.

$$\sigma_{IL} = \sqrt{\sigma_p^2 \cos^2(\theta) + \sigma_p \sigma_t \sin^2(\theta)} \tag{10.13}$$

In the application of Equation 10.13 under the idealized assumption of thin beds with infinite lateral extent, σ_p and σ_t are given by the parallel and series averages, Equations 10.9 and 10.10. Thin beds of finite lateral extent may be modeled using effective medium theory as discussed in Chapter 6.

Laterolog: *Dip effects.* Equation 10.13 does not apply to laterolog measurements. No simple analytic expression has been found for laterolog resistivity in a dipping thin-bedded formation. Numerical simulation of the laterolog in this geometry requires a full 3-D model.

Triaxial induction: *Dip effects.* Multi-component, or triaxial, induction tools may afford the opportunity to measure a component of formation conductivity in the direction parallel to the borehole, in addition to the standard induction conductivity expressed by Equation 10.13. It is fairly easy to use a rotation of the conductivity tensor to show that the conductivity parallel to the borehole can be expressed as in Equation 10.14. We label this component σ_z to refer to an (x, y, z) coordinate system with z-axis parallel to the borehole. The multi-component tool also measures the "parallel" component approximated by Equation 10.13.

$$\sigma_Z = \sigma_p \sin^2(\theta) + \sigma_t \cos^2(\theta) \tag{10.14}$$

Convolution models

Convolution modeling of vertical well-log responses originated with Doll's geometric factor theory for the induction tool [Doll, 1949]. Subsequently Schlumberger introduced what became the industry-standard induction tool, the 6FF40, in 1959 [Tanguy, 1967]. There were many publications on convolution and deconvolution modeling of the 6FF40 accounting for skin effect [Moran, 1982; Gianzero and Anderson, 1982], dip [Howell and Fisher, 1982], and even invasion effects [Meyer, 1987].

Convolution models for a specific tool may be derived by an approximate solution of the tool's response or by numeric modeling. In either case specific parameters of the tool's geometric configuration are required, and for tools in current use these parameters are generally proprietary. Convolution filters for current tools may be obtained from the logging service companies.

Interval-average models

Induction conductivity. Consider the earth model depicted in Figure 10.17(A), where a borehole penetrates a defined thin-bedded interval at relative dip angle θ. The interval-average parallel and transverse conductivites of the earth model across the defined interval ($\sigma_{p,\,INT}$ and $\sigma_{t,INT}$) are defined by Equations 10.9 and 10.10, where the volume fractions are computed across the defined interval. Figure 10.17(B) shows an equivalent homogeneous and transversely isotropic interval with the same parallel and transverse conductivities. Schematic induction conductivity curves (labeled σ_{IL}) are drawn along the borehole in each figure. In Figure 10.17(B), the induction conductivity reaches a constant value in the center of the defined interval which is given by Equation 10.13. This constant value is the *average* of the induction conductivity across the defined interval in Figure 10.17(A).

Figure 10.17. Induction conductivity across defined thin-bed interval (A) and equivalent homogeneous, transversely isotropic interval (B).

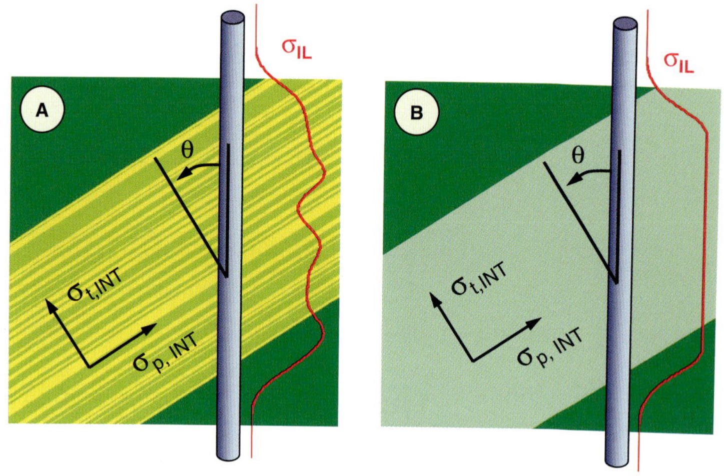

The interval-average model for the induction conductivity is, therefore, given by Equation 10.15, where $\sigma_{IL,INT}$ is the average of the measured induction conductivity across the defined interval.

$$\sigma_{IL,\,INT} = \sqrt{\sigma^2_{p,\,INT}\cos^2(\theta) + \sigma_{p,\,INT}\,\sigma_{t,\,INT}\sin^2(\theta)} \qquad (10.15)$$

The average conductivity, $\sigma_{IL,INT}$, should be estimated in a manner that minimizes the effect of the shoulder beds above and below the defined interval. Equation 10.8 approximates the magnitude of the error caused by the shoulder beds if $\sigma_{IL,INT}$ is computed across all the measure points in the defined interval; see the discussion on page 148.

Deep laterolog. The interval-average model for laterolog resisitivity does not account for dip effects, so it is valid only when the borehole is perpendicular to bedding. In Equation 10.16, $\sigma_{p,\,INT}$ is the parallel conductivity of the earth model as defined by Equation 10.9, where the volume fractions are computed across the defined interval. $\sigma_{LLD,INT}$ is the average conductivity across the defined interval, computed from the measured LLD resistivities.

$$\sigma_{LLD,\,INT} = \sigma_{p,\,INT} \qquad (10.16)$$

Summary

Any mathematical model of a logging tool's measurement requires a geometric and petrophysical description of the measured earth region. This description, termed an *earth model*, comprises a subdivision of the earth region into subregions each having relatively constant petrophysical properties, together with the corresponding property values for each region. An earth model may have one-, two-, or three-dimensional geometry, and it may be *explicit* or *statistical*. In an *explicit* model, the geometry and the properties of each individual subregion (e.g., bed) are defined explicitly. In a *statistical* model no individual beds are defined; instead, the description consists of a statistical distribution of *bed types* and their average properties.

The process of calculating a well-log response numerically for a given earth model is referred to as *forward modeling*. The process of estimating the geometric and petrophysical parameters of an unknown earth model from a set of given field logs and appropriate forward models is referred to as *inverse modeling* or *inversion*. All formation evaluation techniques, whether conventional or exotic, utilize some form of log modeling and inversion.

Forward models for logging tool responses may be classified into four categories: Detailed numeric simulations of tool physics in one, two, or three dimensions; simplified approximate solutions to tool equations; one-dimensional convolution models; and interval-average models.

The two main techniques of inversion for thin-bed applications are iterative numeric inversion, which yields a single (but often non-unique) solution for earth-model properties, and Monte Carlo inversion, which yields probabilistic distributions of earth-model properties.

For lithology and porosity logs in thin-bed applications, the practical forward models are either convolution models or interval-average models. Gamma-ray, acoustic, density, and neutron log models are discussed in this chapter.

All four categories of forward model are employed for resistivity logs in thin-bed applications. Approximate solutions, convolution models, and interval-average models for resistivity tools are discussed in this chapter; detailed numeric simulations are discussed in Chapter 11.

Deep-water–Hemipelagic,
Miocene,
Monterey Formation:
Montana de Oro State Park, California, U.S.A.

Photo by Quinn Passey

11 High-resolution Techniques

Overview

Introduction

This chapter describes in more detail the techniques introduced in Chapter 2 for the high-resolution evaluation of thinly bedded clastic reservoirs. In particular, it describes the general techniques of log convolution modeling and the detailed forward modeling and inversion of resistivity tool responses. Examples are presented for each technique, and their relative advantages and limitations are discussed.

Contents

DOI: 10.1306/1157794A13220

Introduction and Recommendations

Two components

Our approach to high-resolution evaluation has two components, both based on the concepts of log modeling and inversion introduced in Chapter 10. The first component, termed *log convolution modeling* (LCM), uses a high-resolution earth model and convolution filters to invert all the logs that can be considered to have linear responses that are not significantly disturbed by borehole or invasion effects. Logs treated by LCM may include lithology, porosity, and resistivity logs. The second component is *resistivity log modeling* using detailed forward models for resistivity tools to account for two- or three-dimensional and nonlinear effects including combinations of invasion, borehole, and dip effects. A complete evaluation might combine detailed modeling of resistivity logs with LCM for lithology and porosity logs.

High-resolution earth model

By definition, a *high-resolution* evaluation means that each thin bed in the reservoir interval is identified and analyzed individually. Such an evaluation obviously depends on the construction of an accurate *high-resolution earth model*, where bed boundaries are identified by means of a borehole image log or other high-resolution measurement. Limits on image-log resolution suggest that it is not practical to even attempt a high-resolution evaluation where a significant fraction of beds are thinner than 2 in. [5 cm]. As indicated below, other factors raise the minimum workable bed thickness even higher.

The high-resolution evaluation may use LCM or resistivity modeling alone, or both components in combination. The use of a common earth-model geometry for all logs ensures there will be no resolution mismatch artifacts in the final results.

Resistivity is critical

Accurate modeling of the resisitivity log is the most critical element of a successful high-resolution evaluation. The resistivity log has the lowest native resolution of all the conventional log data employed in the evaluation. Further, the resistivity is highly sensitive to thin interbedded shales, and small errors in modeling can have a large effect on calculated water saturation. The sensitivity and resulting uncertainty are especially pronounced when the beds are too thin to even be detected by the resistivity log. For this reason it is recommended that high-resolution evaluations be restricted to cases where most of the beds exceed 1 or 2 ft [.3 or .6 m] in thickness. If a significant fraction of beds are thinner than 3 or 4 ft [1 or 1.2 m], then a complementary low-resolution analysis [Chapter 12] should be performed to help validate the analysis and establish levels of uncertainty in the results.

Advantages and limitations

The primary rationale for performing a high-resolution evaluation of a thinly bedded reservoir has been described in Chapter 1: Conventional log analysis commonly results in significant underestimation of hydrocarbon pore thickness, and a high-resolution evaluation has the potential to produce a much more accurate estimate.

The advantage of the high-resolution approach, compared to the low-resolution method described in Chapter 12, is that a high-resolution evaluation produces a detailed result that identifies each reservoir bed and its properties, displayed in familiar well-log format. In contrast, the low-resolution approach does not identify individual beds, but produces a result that represents only average bed properties and total hydrocarbon pore-thickness (HPT) across the thinly bedded reservoir interval.

The most significant limitation or risk associated with high-resolution evaluations is that well-log inversion does not produce a unique result. The non-uniqueness and associated uncertainty increase as bed thicknesses decrease, and are especially pronounced when beds are thinner than the detection limits of the logging tools. The high-resolution techniques do not provide a built-in mechanism to estimate this uncertainty.

Recommendations

Following is a summary of general recommendations for the proper application of the high-resolution techniques described in this chapter. The sections on LCM and resistivity modeling contain additional recommendations specific to those methods, as well as discussion of their relative advantages and limitations.

1) High-resolution evaluations should mostly be applied in cases where the median bed thickness exceeds 1 or 2 ft [.3 or .6 m].
2) If a significant fraction of beds are thinner than 3 or 4 ft [1 or 1.2 m], use a complementary low-resolution analysis [Chapter 12] to help validate results and establish levels of uncertainty.
3) Use high-resolution data to establish a common earth model geometry for use in modeling all logs in the evaluation. Do not attempt to establish earth-model geometry using data that cannot resolve all the beds.
4) Use all available data to test the validity of any result obtained by a high-resolution inversion of inherently low-resolution well logs.

Log Convolution Modeling

Definition

Log convolution modeling (LCM) is a forward-modeling and inversion process using convolution models for selected logs. First, an initial high-resolution earth model is formulated. The log parameters of the earth model are convolved with one-dimensional approximations of their respective tool response functions and the computed logs are compared with the measured logs. The earth-model parameters are updated and this process is iterated until acceptable agreement is obtained between measured and computed logs.

In other words, LCM combines a 1-D earth model, 1-D convolution models for each well log, and an iterative inversion procedure to obtain an inverted earth model that is taken to be a more accurate representation of true formation parameters than the original logs. The final parameters of the inverted earth model are used to estimate net reservoir thickness, porosity, water saturation, and HPT using a standard Archie or shaly sand analysis in the sand beds. Figure 11.1 illustrates the LCM process by means of a flow chart.

Implementation

Effective implementation of the LCM procedure requires a software application that performs all of the operations depicted in Figure 11.1, up to the construction of the final earth model. Listed below in Table 11.1 are some of the functions that such an application should provide.

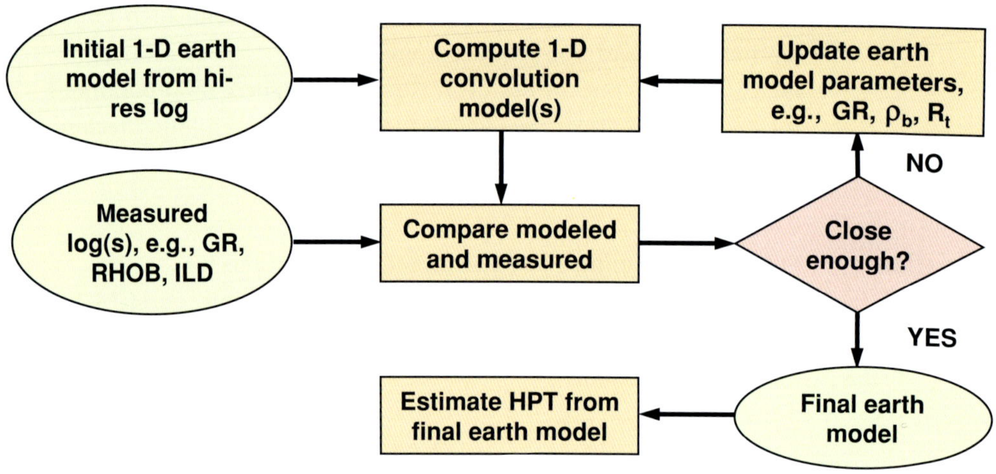

Figure 11.1. Flowchart for log convolution modeling.

Table 11.1. Functions of an application implementing LCM.

Function	Description
Import and export log data	Import and export digital log data in a format compatible with other applications.
Build initial earth model	Block a high-resolution log. Merge boundaries from several blocked logs. Provide interactive graphical editing of bed boundaries and parameter values (e.g., ρ_b, R_t).
Compute 1-D convolution models	Incorporate a library of convolution filters for all commonly used logs.
Compare measured and computed logs	Graphical comparison plus quantitative comparison based on a sum of squared errors.
Update earth model	Interactive graphical editing of earth model, both parameter values and bed boundaries. Graphical and quantitative comparison should be automatically updated with each interactive model change. (Also potentially may incorporate *automatic* parameter updating to minimize sum of squared errors.)
Export final earth model	Export final earth model as blocked log curves in format compatible with standard log analysis applications.

There has been at least one commercial implementation of the LCM process. This was the SHARP™ procedure, patented and performed as a customer service by Schlumberger [Serra and Andreani, 1991]. ExxonMobil uses an independently developed proprietary application to implement LCM.

Cautionary example

Figure 11.2 illustrates rectangular earth models (black), LCM-computed logs (red dashed), and measured log data (blue solid) for a 50-ft [15-m] interval of the gas-bearing South Texas reservoir. The sandstones are characterized by low permeability (from less than 0.1 mD to 10 mD) and shallow invasion.

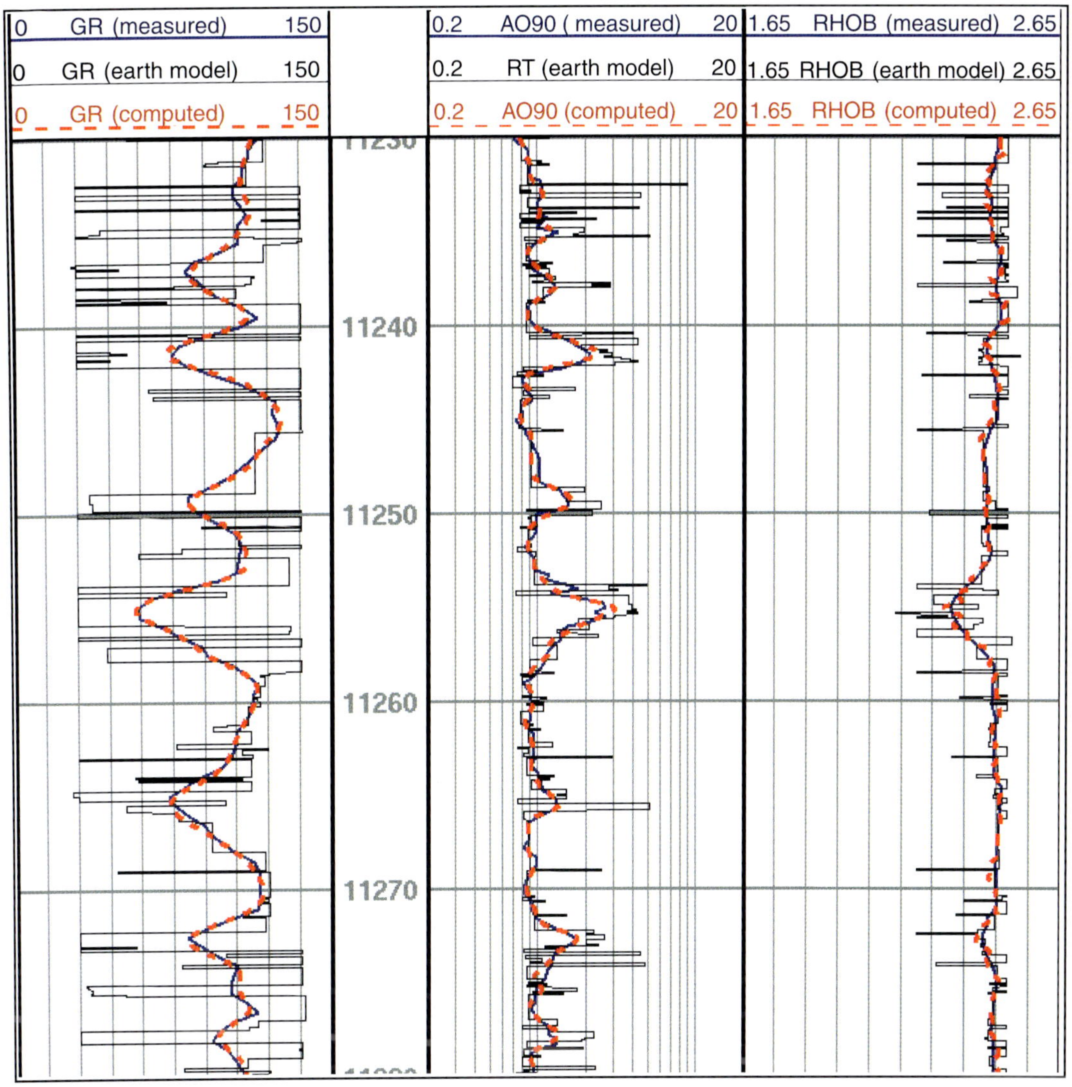

Figure 11.2. LCM modeling results for thinly bedded lower-shoreface to prodelta deposits, Vicksburg formation.

In this example, high-quality borehole image data from a near-gauge hole give high confidence in bed boundary locations and relative resistivity response, and good agreement is obtained between the measured and computed logs.

In the Vicksburg formation, thin sandstones appear bright (resistive) relative to more clay-rich "shales" or siltstones on borehole image (EBI) logs. Figure 11.3 shows a close-up of a 2-ft [.6-m] section of the modeled interval with the EBI image data. Note that the GR, R_t, and porosity earth models are in stark contrast to the flat response of the measured logs through the interval. Assuming the earth model is correct, this interval would yield 20% net-to-gross, 22–27% porosity, and 55–75% gas saturation after LCM modeling. In contrast, the conventional log analysis indicates no net sand or net pay.

The Vicksburg result shown here exemplifies good modeling practices using high-quality data. *Unfortunately, it also demonstrates the potential for an incorrect solution in the absence of external*

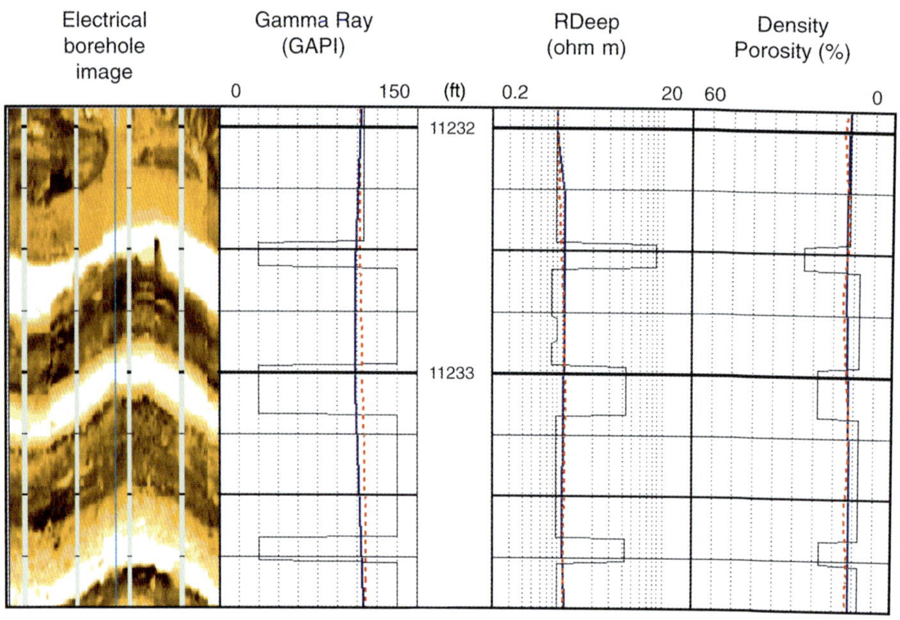

Figure 11.3. Close-up of 2-ft [.6-m] interval of the gas-bearing Vicksburg formation with LCM modeling results (squared logs).

validation. Routine core analysis data that became available after the LCM model was completed revealed that the resistive thin beds in Figure 11.3 (and elsewhere in the interval) were tight (i.e., very low porosity) and, therefore, probably not gas-bearing.

This example illustrates the inherent uncertainty in high-resolution modeling with beds only a few inches thick. *A good match between measured and computed logs is not proof of the model's validity.*

Advantages and limitations

ADVANTAGES:

- Log convolution modeling provides a means to resolution-match all the logs employed in a thin-bed evaluation, including lithology and porosity as well as the resistivity logs. As was shown in Chapter 1, failure to resolution-match the porosity log can have a significant negative impact on HPT when bed thicknesses are in the range of 2 ft [.6 m] or less.
- LCM is generally more rapid to apply than detailed resistivity modeling. The computation times for detailed forward models may be substantial, while convolution models are extremely fast to compute.

LIMITATIONS:

- LCM should not be applied to resistivity logs when borehole, invasion or dip effects are significant, or when induction resistivity is outside the "linear range" of 1–20 ohm m.
- LCM should not be applied to logs whose response characteristics are strongly non-linear, such as spontaneous potential and old unfocused electric logs.
- Convolution models for porosity tools are approximate and are based on highly simplified assumptions about the functioning of these tools. The effect of these approximations on the uncertainty in the results has not been thoroughly investigated.
- As with any high-resolution technique, LCM does not produce a unique solution. The non-uniqueness and uncertainty increase when bed thicknesses are below the detection limit of the logging tools. The example above (Figure 11.3) shows that this non-uniqueness can produce misleading results even when reasonable care is applied in the analysis.

Recommendations

When to apply: In considering the application of LCM, the following questions should be addressed. Affirmative responses to these questions support the application of LCM.

1) Are high-resolution logs available (e.g., EBI or dipmeter), and do they accurately identify bed boundaries?
2) Is invasion negligible? 1-D convolution modeling does not account for invasion or borehole effects.
3) Are induction log resistivities in the range of 1–20 ohm m? Outside this range, nonlinearities in induction tool response limit the accuracy of the convolution modeling approach and detailed resistivity tool modeling should be considered.
4) Are there core data, or log data from thicker reservoir sandstones and shales, to help constrain selection of earth-model parameters such as R_{sh}, R_T, and porosity?

LCM procedure:

- Ensure that all log and core data are carefully depth-aligned before beginning analysis.
- Perform borehole corrections to logs before applying LCM.
- Where possible, perform quantitative comparisons of the high-resolution log(s) used for bed definition with core image data to evaluate the accuracy of bed-boundary picks.
- Use all available data, including core data and logs from thicker intervals, to constrain the selection of earth-model parameters in both reservoir and non-reservoir beds types.
- For an accurate reservoir evaluation, the earth-model parameters in the non-reservoir beds are equally as significant as those in the reservoir beds. Relatively small variations in shale parameters, for example, can often have a large impact on the parameters required in the reservoir beds to produce a match between measured and computed logs.
- Earth-model parameters for each non-reservoir bed type should be held relatively constant and should be validated with other data.

Uncertainty and validation:

- Generate best-case, worst-case, and most-likely LCM inverse solutions for the field data.
- A low-resolution analysis (Chapter 12) can help validate the earth-model parameters and the final results of an LCM analysis, and can help assess the uncertainty in the results.

Resistivity Modeling

Definition

Resistivity modeling involves replicating an observed field log by numerically solving the mathematical boundary value problem of the electromagnetic fields generated from a specific resistivity tool for a pre-defined layered earth model. To the degree that the field log and the modeled log responses are in acceptable agreement, the underlying earth model can be considered one possible representation of the formation's true resistivity profile.

Thus, resistivity modeling is a process that combines an initial earth model which may be one-, two-, or even three-dimensional with detailed numeric simulations of one or more resistivity tools in an iterative inversion procedure to obtain an inverted earth model that is taken to be a more accurate representation of true formation resistivity than the original logs. For a complete evaluation of hydrocarbon pore-thickness in a thinly bedded reservoir, a detailed resistivity model may be combined with convolution models for lithology and porosity logs using earth models with a common set of bed boundaries.

The dimensionality of earth models is discussed in general terms in Chapter 10. The specific resistivity-log effects addressed by different dimensions of earth model are summarized in Table 11.2. The focus in this chapter will be on 1-D and 2-D modeling.

Table 11.2. Resistivity effects addressed by modeling in 1-D, 2-D, 3-D.

Dimension	Effect addressed
1-D	Thin beds Shoulder beds Dipping beds
2-D	Borehole Invasion Anisotropy (transverse isotropy)
3-D	Heterogeneous resistivity. Examples: Sand injections; non-planar bedding geometry; combined dip, borehole and invasion effects.

Implementation

Resistivity modeling requires a specialized software application that enables the petrophysicist to build and modify earth models, compute resistivity forward models, and compare computed and measured logs. Listed below in Table 11.3 are the basic functions that such an application must provide. Figure 11.4 illustrates the 2-D earth model features that the application must be able to describe in addition to the depths of bed boundaries.

2-D example

The 4-m [13-ft] log section plotted in Figure 11.5 is situated a few m above the well's main target, a 60-m [197-ft] gas reservoir interval. The gamma ray (GR) indicates a sand about a meter thick at XX54. The density and neutron porosity curves show apparent gas crossover at this depth (note the non-standard scales). The deep laterolog (LLD) reads only about 4 to 5 ohm m in the sand, while the two logging-while-drilling (LWD) resistivities, R-BIT and R-RING, read 10 and 20 ohm m respectively.

The discrepancy between the LLD and LWD resistivities, together with separations between the deep and shallow laterologs (LLD and LLS) in this interval and in the main reservoir, led the analyst to suspect that the LLD was affected by a combination of shoulder-bed and invasion effects. A 2-D resistivity modeling study was performed to resolve this question and obtain improved resistivity estimates for the thin sand illustrated in Figure 11.5 and for the main reservoir. A first-pass earth model was constructed by blocking a resistivity curve derived from an electrical borehole image log (EBI). This earth model is plotted in Figure 11.5 as the red-squared resistivity, RTSQ.

Table 11.3. Functions of a resistivity modeling application.

Function	Description
Import and export log data	Import and export digital log data in a format compatible with other applications.
Build initial earth model	Provide facility for constructing 1-D and 2-D earth models, including varying borehole size, invasion profiles, and dipping beds (Figure 11.4).
Compute forward models	Incorporate a library of forward modeling codes for all resistivity tools of interest.
Compare measured and computed logs	Graphical comparison plus quantitative comparison based on a sum of squared errors.
Update earth model	Interactive graphical editing of earth model including bed boundaries, invasion depths, and parameter values.
Export final earth model	Export final earth model in format compatible with LCM application and standard log analysis applications.

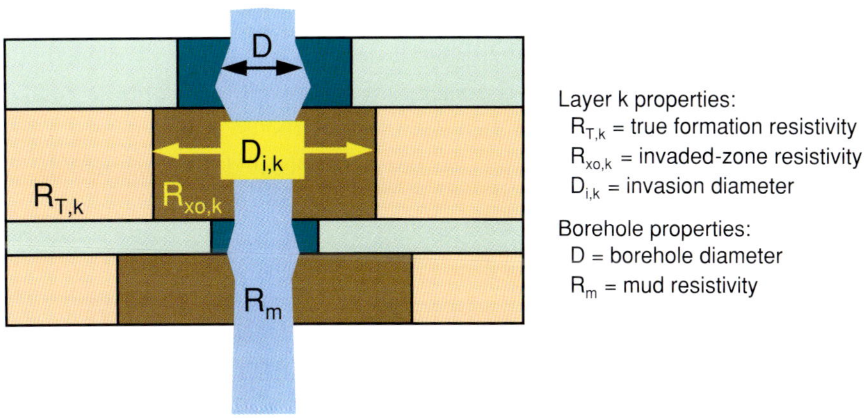

Figure 11.4. Earth model features for 2-D resistivity modeling.

Before inverting the measured LLD, LLS, and MSFL (micro-spherically focused log) resistivities, a benchmark forward-modeling study was performed to investigate the behavior of the deep and shallow laterolog in a 3-ft [1-m] sand bed with and without invasion. This study, illustrated in Figure 11.6, used two hypothetical 3-ft [1-m] sand beds, one with no invasion and one with a 15-in. [38-cm] invasion radius. For the benchmark, the invaded-bed R_{xo} (1.2 ohm m), the shoulder-bed resistivity (1.5 ohm m), the borehole diameter (12.5 in. [31.8 cm]), and the mud resistivity (.092 ohm m) were obtained from the well data. Both sand beds were assigned a true resistivity of 20 ohm m.

The modeled 3-ft [1-m] lower sand in Figure 11.6 shows that, in the absence of invasion, the deep laterolog resolves the true resistivity. But in the upper sand with 15 in. [38 cm] of salty mud invasion, the LLD is reduced to a reading of only about 2.5 ohm m.

From this benchmark study it was concluded that the LWD resistivities are closer to the true resistivity in Figure 11.5, while the LLD is affected by invasion; and that 2-D modeling and inversion of the laterolog data should produce a more accurate resistivity profile for the whole reservoir.

The process of inverting the laterolog data began with adjusting the bed boundaries in the initial earth model (RTSQ in Figure 11.5) to make them consistent with the laterolog, gamma ray, density and neutron logs. With bed boundaries established, the iterative inversion procedure was carried out. The goal in the iterative adjustment of sand-bed resistivities and invasion depths was to match the computed logs to the field logs with a relative error of 5% or less. Six iterations were required. The logs modeled were the deep and shallow laterolog, LLD and LLS. The micro-spherically focused log (MSFL) was used to establish the value for R_{xo} which was held fixed through the process. Figure 11.7 shows the result for the 1-m [3-ft] sand first illustrated in Figure 11.5.

The right track in Figure 11.7 shows the result of the inversion. The final resistivity earth model is the red curve, RTSQ, which indicates 40 ohm m for the thin sand at XX54 m. The modeled LLD is the dotted blue curve, which is in close agreement with the measured LLD shown in solid blue.

For the thin sand in Figure 11.7, the estimated water saturation was reduced from 50%, based on the original LLD, to 20% based on the inverted RTSQ. In the main reservoir (not shown) the modeled resistivity yielded a 15% increase in average gas saturation and a 28% increase in hydrocarbon pore-thickness, which translates directly to a 28% increase in gas-in-place.

This example illustrates the value of detailed 2-D resistivity modeling to estimate R_T when measured logs are affected by shoulder beds and invasion. In contrast traditional chartbook corrections are difficult to apply, and each chart is constructed to be valid for a specific set of formation and borehole conditions, which is usually not duplicated in actual field data.

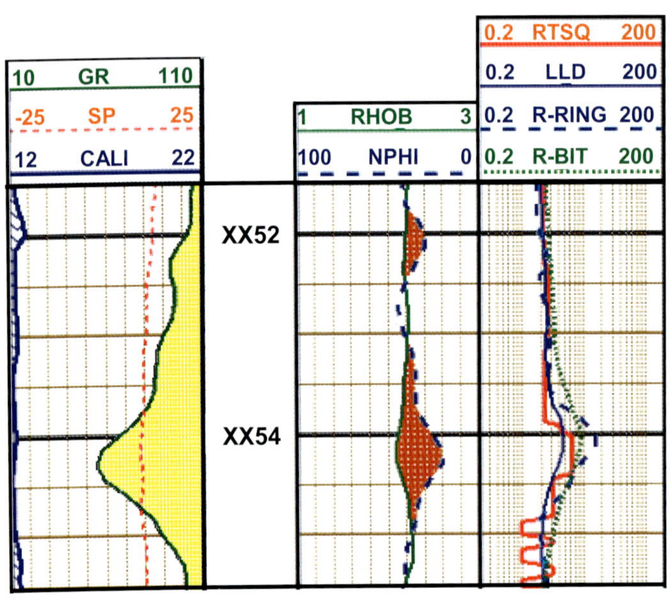

Figure 11.5. LWD and wireline logs across a 1-m [3-ft] sand.

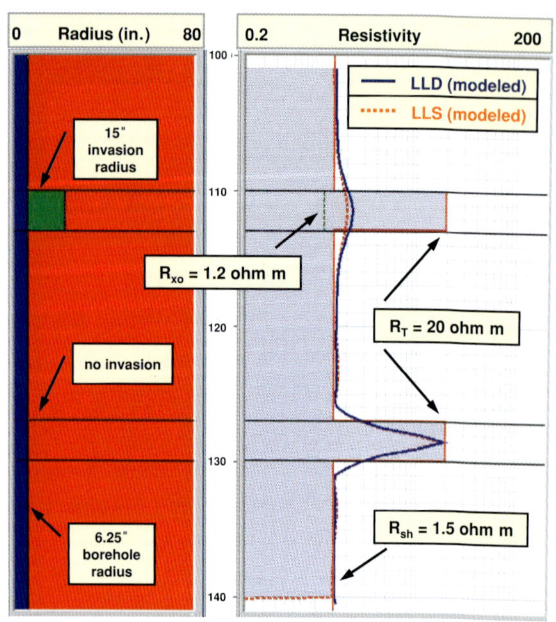

Figure 11.6. Benchmark forward-modeling study for laterolog behavior in 3-ft [1-m] sands.

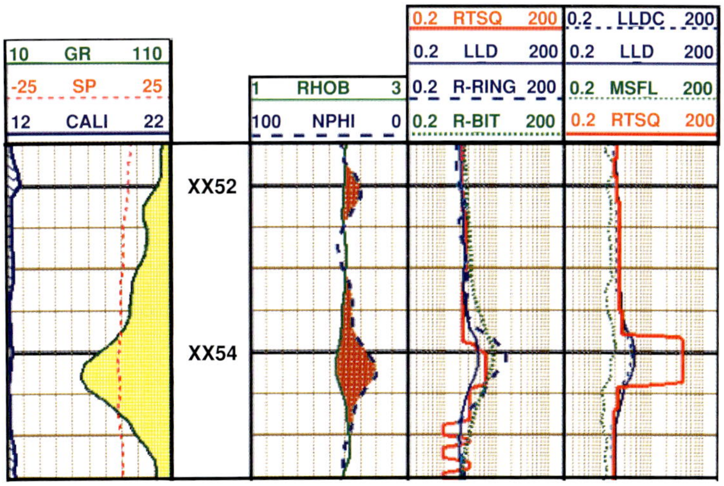

Figure 11.7. Result of 2-D inversion of the laterolog, accounting for shoulder bed and invasion effects. The final modeled resistivity is RTSQ in the right track.

The preliminary benchmark study (Figure 11.6) added credibility to the inversion results and was a quick way to test the feasibility of the proposed interpretation before performing the complete log inversion, which was fairly time-consuming.

Non-uniqueness and uncertainty

For a given earth model and resistivity-tool forward modeling algorithm, there is a unique solution for the resistivity tool's response to the postulated earth model. However, since both the earth model and the modeling algorithm are idealized approximations of reality, there is always some error in comparing a modeled log to a measured resistivity log.

Conversely, given measured resistivity log data and forward modeling algorithms for the resistivity tools, there is no unique inverse solution for the earth model that describes the logged formation. There are several reasons for this *non-uniqueness*:

- The modeled logs can only be matched to the measured logs to within a specified tolerance. Some range of earth-model parameters will satisfy this tolerance. The appropriate tolerance is related to the accuracies of the tools and of the forward models.
- The range of earth-model parameters that produces an acceptable match depends on the sensitivity of the particular tools to bed thicknesses, resistivities, invasion depths, etc.
- If there are beds thinner than the tools' resolution or detection limits, then the resistivity logs alone cannot define the locations or thicknesses of these beds. In this case, numerous postulated thinly bedded earth models may produce an acceptable match.

The *non-uniqueness* of inversion results in *uncertainty* in the results of a high-resolution evaluation of a thinly bedded reservoir. Resistivity modeling and inversion is a complex procedure and there is no direct, general, and complete way to estimate this uncertainty. One case-specific approach is to constuct earth models representing best, most-likely, and worst-case solutions for a given evaluation. A useful general approach is to perform simplified benchmark forward-modeling studies that provide insight into the sensitivities of specific resistivity tools under a particular set of logging conditions.

The following pages present several such benchmark examples for the classic 6FF40 induction tool. The purpose of these examples is twofold: First, to show how benchmark modeling studies can be utilized to gain insight into log response in thin beds; and second, to provide some general insight into the behavior of the induction tool. The examples use particular sets of earth-model parameters, and the specific numeric results should not be generalized to situations with different parameters.

The first two examples investigate the sensitivity of the induction log to small changes in bed thickness and bed resistivity when the beds are just thick enough to be resolved. In both cases we start with 6-ft [2-m] sand and shale beds, where the shale beds are 1 ohm m and the sand beds are 20 ohm m.

Figure 11.8 illustrates the sensitivity of the induction log to 5% changes in bed thickness while the bed resistivities are held constant.

Figure 11.8 shows that, when the thicknesses of the resistive and conductive beds are changed 5% while the bed resistivities are held constant, the calculated induction resistivity remains almost constant in the conductive beds but changes about 15% in the resistive beds. This shows that *induction response in resistive beds is very sensitive to bed thickness. Therefore, accurately defined bed thickness is critical in deriving accurate R_T for resistive beds in the inversion of an induction log.*

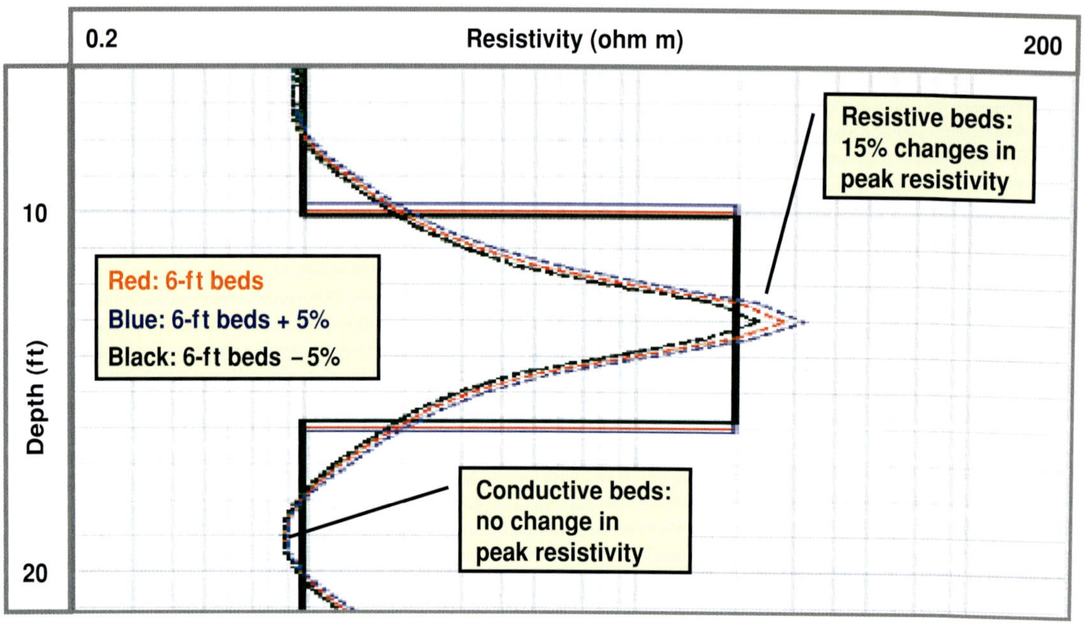

Figure 11.8. Induction log response to 5% changes in bed thickness for 6-ft [2-m] beds of unchanging resistivities, 1 and 20 ohm m.

Figure 11.9 illustrates the sensitivity of the induction log to 5% changes in bed resistivity while the bed thicknesses are held constant.

Figure 11.9 shows that, when the resistivities of the resistive and conductive beds are changed 5% while the bed thicknesses are held constant, the calculated induction resistivity changes 5% in the conductive beds but remains almost unchanged in the resistive beds. This shows that *for 6-ft [2-m] beds, the induction response is more sensitive to relative resistivity changes in the conductive beds than*

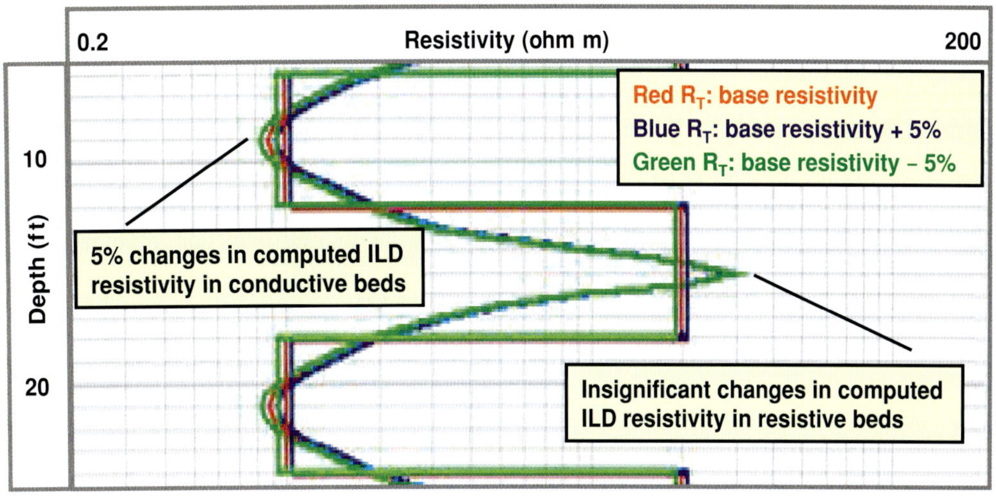

Figure 11.9. Induction log response to 5% changes in bed resistivity for 6-ft [2-m] beds of unchanging thickness.

in the resistive beds. See Figure 11.11 below for an example of this sensitivity difference in a thinly bedded formation.

The next two examples investigate the sensitivity of the induction log to changes in bed resistivity when the beds are thinner than the induction tool's detection limit. Both examples use a sequence of 1-ft [.3-m] beds.

Figure 11.10 illustrates the induction response when the shale beds are fixed at 1 ohm m and the sand beds range from 10 to 100 to 1000 ohm m. Both the earth model and the computed log for 10 ohm m case are plotted in blue. The 100 ohm m case is plotted in black, and the 1000 ohm m case is in red. Note that the computed responses for the 100 ohm m and the 1000 ohm m cases cannot be differentiated on the log plot.

The response illustrated in Figure 11.10 can be approximated by the familiar parallel conductivity equation, repeated here as Equation 11.1.

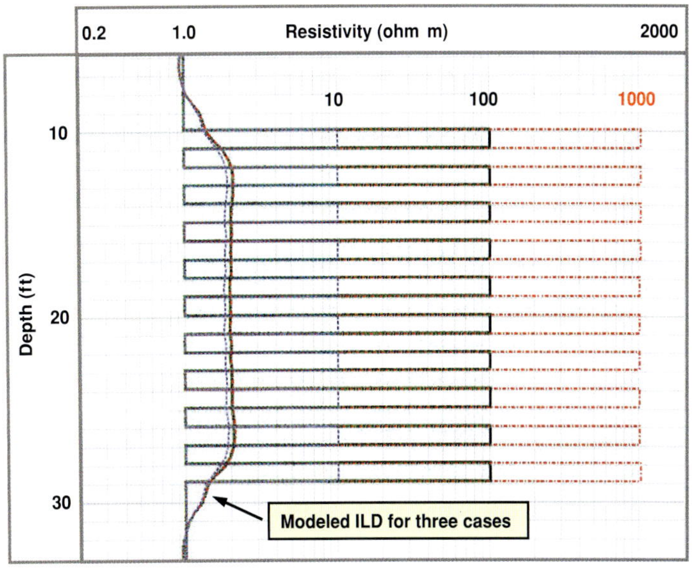

Figure 11.10. Modeled deep induction log (ILD) response in 1-ft [.3-m] beds with sand resistivities of 10, 100, and 1000 ohm m.

$$\sigma_{ILD} = v_{sh}\sigma_{sh} + (1 - v_{sh})\sigma_{sd} \tag{11.1}$$

In Equation 11.1 σ_{ILD} is the induction conductivity, σ_{sh} is the shale conductivity, σ_{sd} is the sand conductivity, and v_{sh} is the shale fraction. Table 11.4 summarizes parameters for Figure 11.10, where $v_{sh} = 0.50$.

Table 11.4. Parallel conductivity approximation for induction response.

R(shale) (ohm m)	R(sand) (ohm m)	σ_{ILD} (mS/m)	R_{ILD} (ohm m)
1	10	550	1.818
1	100	505	1.980
1	1000	500.5	1.998

The accuracy of induction tools is usually cited in terms of conductivity, and a typical accuracy for the range illustrated in Table 11.1 is 2%, or about 10 mS/m. Note that the difference in induction response between the 100 ohm m and 1000 ohm m cases is only 4.5 mS/m! Thus, we see that this difference is too small to be detected quantitatively by the induction tool. *The standard induction-log measurement in a thinly bedded formation like that illustrated in Figure 11.10 is inherently unable to distinguish between sand beds of 100 ohm m or 1000 ohm m.*

Figure 11.11 illustrates the induction response for two sets of interbedded 1-ft [.3-m] sandstones and shales. For the black earth model and black log response, the shales are 1 ohm m and the sands are 100 ohm m. For the red earth model and log response, the shales are 1.1 ohm m and the sands are 9.91 ohm m. Note that the log responses are indistinguishable. Evidently, in a thinly bedded formation, the induction response is much more sensitive to changes in the shale resistivity than in the sand resistivity. Like the previous example, this conclusion can be understood using the simple parallel-conductivity relationship, Equation 11.1.

Since the two log responses in Figure 11.11 are indistinguishable, this means that either the black or the red earth model would be a valid result when inverting a measured log like that shown. This

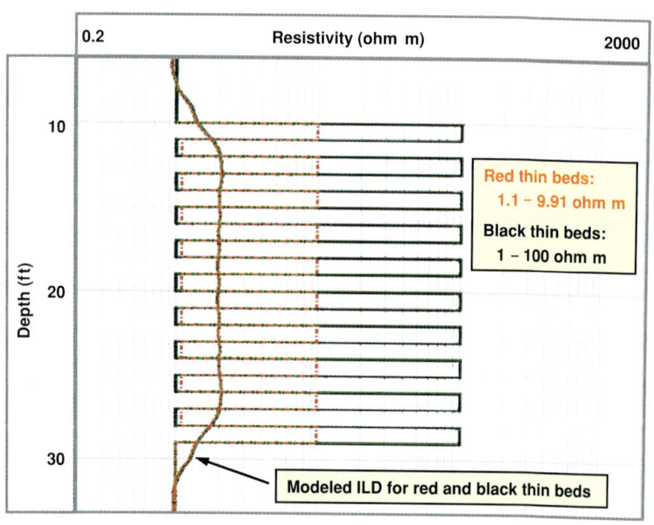

Figure 11.11. Modeled deep induction response for two sets of sand and shale resistivities.

observation can also be expressed as follows. *In inverse modeling the induction log for a formation like Figure 11.11, a 10% error in shale resistivity can produce a 90% error in sandstone resistivity.*

The final example of non-uniqueness in induction response addresses what can happen if high-resolution data are not available to establish bed boundaries for inverse modeling. Figure 11.12 shows modeled induction logs for two cases. The earth model on the left comprises interbedded 1-ft [.3-m] shales of 1 ohm m and sandstones of 20 ohm m. The model on the right is a 19-ft [6-m] bed of 1.9 ohm m with 1 ohm m shoulder beds. The induction responses are essentially identical; the difference in conductivities at the center is smaller than the tool's accuracy limit. Thus, *in the absence of high-resolution data to establish bed boundaries, either earth model in Figure 11.12 is a valid result for inverting the log response shown.*

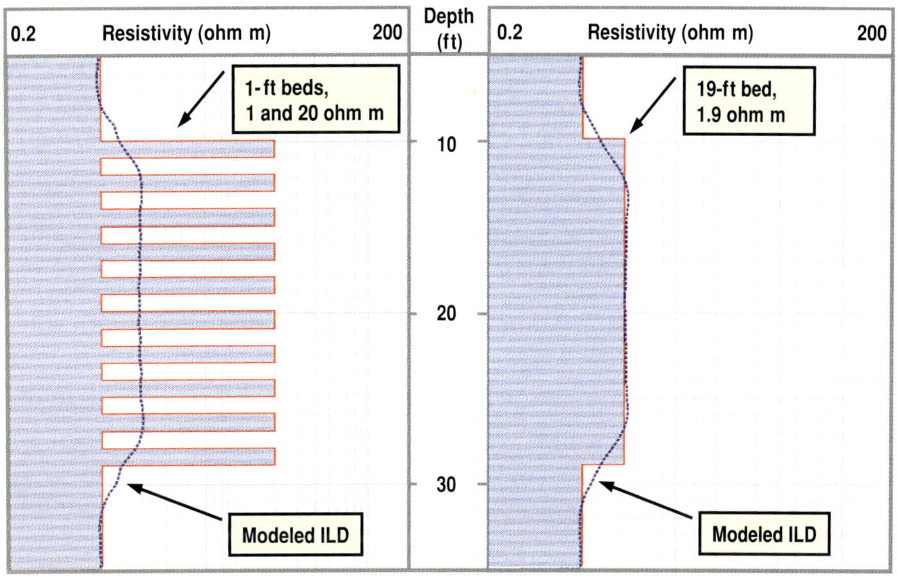

Figure 11.12. Near-identical induction responses to a sequence of thin beds and a single thick bed.

Advantages and limitations

ADVANTAGES:

- Resistivity modeling and inversion can significantly improve the accuracy of HPT evaluations in thinly bedded reservoirs.
- Detailed resistivity modeling (as contrasted to convolution modeling) is particularly useful when the resistivity log is affected by invasion, rugose borehole, dipping beds, or other factors that cause non-linear tool responses.
- Service-company correction charts for invasion, borehole corrections, shoulder-bed corrections, etc., are difficult to apply and are only valid for the specific conditions that were used in designing the charts. Tool modeling and inversion, in contrast, can handle any combination of borehole and invasion conditions.

LIMITATIONS:

- Inverse modeling of resistivity logs often does not produce a unique solution. This non-uniqueness is especially pronounced in thinly bedded formations where many beds are thinner than the resistivity tool's detection limit.

- Auxiliary high-resolution data are required to define bed boundaries and reduce the uncertainty associated with the non-uniqueness of resistivity inversion.
- The computational requirements of detailed forward-modeling algorithms for resistivity tools can make the process of resistivity inversion fairly time-consuming.
- Because of the computational requirements of the forward-modeling algorithms, there is no general way to evaluate the uncertainty associated with the non-uniqueness of resistivity inversion.

Recommendations

When to apply:

- High-resolution evaluations should be applied mostly in cases where median bed thickness exceeds 1 or 2-ft [.3 or .6-m].
- Detailed resistivity modeling should be considered when the resistivity log is affected by invasion, rugose borehole, dipping beds, or other factors that cause non-linear tool responses.

Inverse modeling procedure:

- Use all available petrophysical and geological data to constrain the earth model used for inversion. Examples:
  + Define bed boundaries using high-resolution data such as borehole images, other high-resolution logs, core images, or core descriptions.
  + Use log measurements in thick beds to constrain the resistivities for thin sand and shale beds.
  + Use shallow resistivity logs to constrain R_{xo} in invaded beds.
  + Limit invasion depth in shales to zero.
- The iterative inversion process should be stopped when the difference between the computed and field logs is within a reasonable limit. This limit may be inferred from the tool's specified measurement accuracy. For galvanic resistivity tools, the measurement accuracy is usually defined as ±5% of the measured apparent resistivity. For induction tools, the measurement accuracy is defined in terms of conductivity, ±2% or 0.75 mS/m (at 20 kHz), whichever is higher.
- When a reasonable match between the field and computed logs cannot be obtained within ten or fewer iterations, the inversion process should not be carried further without checking the validity of the assumptions in the earth model.

Uncertainty and validation:

- Use benchmark forward-modeling studies to evaluate the range of earth models that might produce log responses similar to the field data.
- Generate best-case, worst-case, and most-likely inverse solutions for the field data.
- If appropriate, perform a parallel low-resolution analysis [Chapter 12] that can provide insight about the uncertainty in the HPT evaluation.

Summary

Our approach to high-resolution evaluation has two components. The first, termed *log convolution modeling* (LCM), uses a high-resolution earth model and convolution filters to invert all the logs that have linear responses and are not significantly disturbed by borehole or invasion effects. Logs treated by LCM may include lithology, porosity, and resistivity logs. The second component is *resistivity log modeling* using detailed forward models for resistivity tools to account for two- or three-dimensional and nonlinear effects including combinations of invasion, borehole, and dip effects. A complete evaluation might combine detailed modeling of resistivity logs with LCM for lithology and porosity logs.

Accurate modeling of the resisitivity log is usually the most critical element of a successful high-resolution evaluation. Resistivity modeling and inversion, combined with LCM as necessary, can significantly improve the accuracy of HPT evaluations in thinly bedded reservoirs.

The advantage of the high-resolution approach, compared to the low-resolution approach described in Chapter 12, is that a high-resolution evaluation produces a detailed result that identifies each reservoir bed and its properties. The main limitation associated with high-resolution evaluations is that well-log inversion does not usually produce a unique result, and the consequent uncertainty is not always addressed.

Both LCM and resistivity modeling require the construction of a high-resolution earth model, which usually relies on an electrical borehole image (EBI) log to identify beds. Accurate identification of bed boundaries reduces some of the non-uniqueness inherent in the inversion process.

To further reduce uncertainty, all relevant auxiliary data should be incorporated into the modeling and inversion procedure. For example, core data or log data from thicker beds may be used to constrain the modeled values of petrophysical parameters in thin reservoir and non-reservoir beds.

A useful general approach to the evaluation of uncertainty in the high-resolution approach is to perform simplified benchmark forward-modeling studies that can provide insight into the sensitivities of a specific resistivity tool under a given set of logging conditions. Examples of such benchmark studies are presented in this chapter.

The low-resolution approach described in Chapter 12 is a complementary method that can provide validation of results and additional insight about uncertainties.

Delta Front and Stream Mouth Bar,
Cretaceous,
Ferron Formation:
Miller Canyon, Castle Valley, Utah, U.S.A.

Photo by Anthony Sprague

12 Low-resolution Techniques

Overview

Introduction

This chapter describes in more detail the techniques introduced in Chapter 2 for the low-resolution evaluation of thinly bedded siliciclastic reservoirs. In particular, it describes and illustrates the method of Volumetric Laminated Sand Analysis. It also describes the application of two special low-resolution logs in evaluating thin-bedded reservoirs. These are the Nuclear Magnetic Resonance and the multi-component induction logs.

Contents

DOI: 10.1306/1157795A13220

Introduction

Low-resolution evaluation

Generically, *low-resolution evaluation* refers to any method of evaluating thinly bedded reservoirs that does not attempt to resolve the properties of each individual thin bed through forward modeling and inversion. We can distinguish a few different categories of low-resolution analysis technique.

1) Conventional log analysis performs calculations at each logged depth as if the log measurements represented the true properties of the formation at that depth. This approach does nothing to account for the effects of thin beds.

2) Variants of the conventional laminated sand analysis [Poupon et al., 1954], which use standard or high-resolution logs to solve the parallel-conductivity equation for sand-bed resistivity. The aim is to improve the calculation of sand-bed water saturation. The shortcomings of this approach are discussed in Chapter 1.

3) A new method employing conventional logs and Monte Carlo inversion of statistical earth models is detailed in this chapter. This method, called Volumetric Laminated Sand Analysis (VLSA), is designed to overcome some of the shortcomings of earlier methods and to deal explicitly with the uncertainty inherent in thin-bed evaluation.

4) Certain non-standard low-resolution logs make measurements that are directly and strongly influenced by hydrocarbons in thinly bedded reservoirs. The main examples are nuclear magnetic resonance and multi-component induction logs. Measurements from these logs can be incorporated to improve the results of conventional log analysis approaches or VLSA. These logs and their application are discussed briefly in this chapter.

Advantages and limitations

The advantage of a low-resolution technique designed to accurately address the thin-bed issue, in comparison to a high-resolution approach, is that it obviates the requirement to precisely define the boundaries of each thin bed in the reservoir. As we saw in Chapter 1, there are many cases where such precise bed definition is impossible. Often the high-resolution approach is very labor-intensive; furthermore it does not directly address the uncertainties inherent in thin-bed evaluation.

The main limitation of low-resolution techniques is that they do not produce a detailed earth model of reservoir bed geometry.

Volumetric Laminated Sand Analysis

Definition

Here is a concise definition of VLSA using concepts introduced in Chapter 10. Figure 12.1 illustrates the definition in a flowchart, and the following section provides more detail.

The *Volumetric Laminated Sand Analysis* (*VLSA*) procedure begins with an *a priori* statistical earth model that describes the potential range of petrophysical properties for each bed type, before the introduction of well-log information. Interval-average forward models are used to predict interval-average well-log response for each instance of the statistical earth model. Monte Carlo inversion of these models is employed to update the statistical earth model, generating final distributions of petrophysical properties that are consistent with the well-log data. These final distributions yield a solution for reservoir hydrocarbon pore-thickness (HPT) plus a description of the uncertainty in the solution.

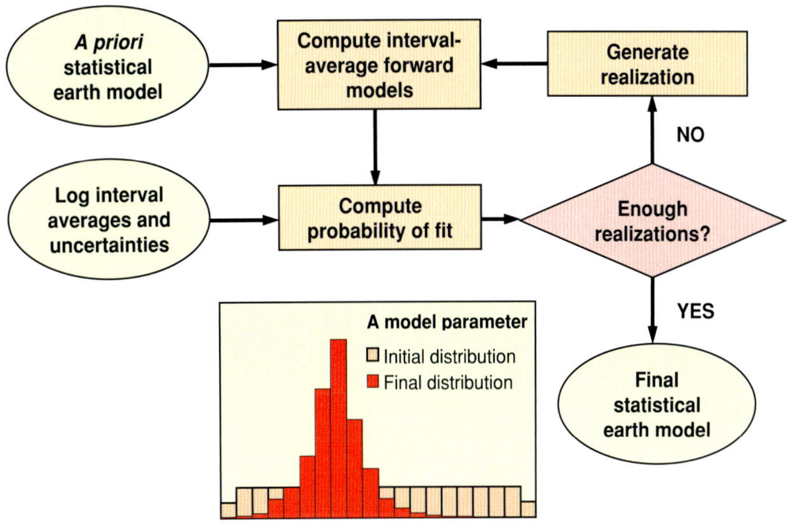

Figure 12.1. Flowchart of VLSA procedure.

VLSA Details

A priori statistical earth model. The initial earth model includes an *a priori* distribution for every petrophysical parameter of each bed type. These *a priori* distributions contain all the information that is known about the parameters before the introduction of well-log data. For example, the information that the average sandstone bed porosity probably lies between 15% and 28% can be formulated as a uniform distribution bounded by those two values. Alternatively, core data might make it possible to describe the mean sandstone porosity by a normal distribution with mean value 24% and standard deviation 1% — a much more tightly constrained initial distribution.

In the Monte Carlo inversion, random samples (*realizations*) are drawn from these *a priori* distributions and tested for consistency with the well-log data. Therefore the final solution is determined not only by the well-log data, but also by the *a priori* knowledge that is incorporated in the initial distributions. This incorporation of *a priori* knowledge is one of the strengths of VLSA.

Interval-average forward models were discussed in general terms in Chapter 10. Here we give two examples to illustrate how the equations of conventional log analysis are incorporated into these models.

First consider the bulk density log for an earth model with two bed types: sandstone and shale. For the interval-average model in Equation 12.1 (based on Equation 10.3), D_{int} is the average of the bulk density log over the analysis interval; v_{sd} and v_{sh} are the volume fractions of sandstone and shale; and ρ_{sd} and ρ_{sh} are the bulk densities of sandstone and shale, respectively.

$$D_{int} = v_{sd}\rho_{sd} + v_{sh}\rho_{sh} \tag{12.1}$$

Now we bring the familiar density porosity equation into the model. In Equation 12.2, ϕ is sandstone porosity, ρ_{ma} is sandstone matrix density, and ρ_{fl} is sandstone fluid density. A similar equation describes shale bulk density in terms of shale porosity and matrix and fluid densities.

$$\rho_{sd} = (1 - \phi)\rho_{ma} + \phi\rho_{fl} \tag{12.2}$$

For this simple two-component earth model, the complete interval-average forward model for bulk density is described by eight petrophysical parameters: The volume fractions, total porosities, matrix densities, and fluid densities for sandstone and shale. Since the sandstone and shale volume fractions sum to 1, the number of independent parameters is only seven:

$$v_{sd} + v_{sh} = 1 \tag{12.3}$$

Next consider the interval-average model for induction log conductivity, again for the simple earth model with two bed types. Assuming a borehole perpendicular to formation bedding planes, the interval-average conductivity model is given by Equation 12.4 (based on Equation 10.9). In Equation 12.4, σ_{IL} is average induction log conductivity across the analysis interval; v_{sd} and v_{sh} are volume fractions; and σ_{sd} and σ_{sh} are the average conductivities of sandstone and shale bed types.

$$\sigma_{IL} = v_{sd}\sigma_{sd} + v_{sh}\sigma_{sh} \tag{12.4}$$

To complete the interval-average model we need to describe each bed type's conductivity in terms of its porosity, water saturation, and electrical parameters. This can be done using any number of standard water saturation equations such as the Archie, Waxman-Smits, or Dual-water models. For simplicity we illustrate here with the Archie model. In Equation 12.5, σ_w is connate water conductivity, ϕ is sandstone porosity, m is the Archie cementation exponent, S_w is sandstone water saturation, and n is the Archie saturation exponent. We can write a similar equation for σ_{sh}, with the simplifying *a priori* information that S_w=1 and that n can be held constant in the shale beds.

$$\sigma_{sd} = \sigma_w \cdot \phi^m \cdot S_w^n \tag{12.5}$$

Interval-average models are constructed similarly for each log employed in the analysis. All of these models use a common set of earth-model parameters including the volume fractions and porosities of each bed type, and each model also has its own specific parameters, such as the matrix and fluid densities for the bulk-density model.

Log interval averages and uncertainties. Log inputs to VLSA are the *averages* of log measurements over the thinly bedded analysis interval; individual depth readings are not used. In addition, the *uncertainties* in these interval averages are an essential input for the computation of the "probability of fit" of each earth-model realization. These interval-average uncertainties depend on both the accuracy and the precision, or repeatability, of the log measurement, roughly as described by Equation 12.6. Here u, a, and r are respectively the interval-average uncertainty, the measurement accuracy, and the pointwise measurement precision of the log, and N is the number of independent measurements averaged over the analysis interval.

$$u^2 = a^2 + r^2/N \tag{12.6}$$

Probability of fit. When the interval-average forward models are computed for a new realization of the statistical earth model, there is a finite probability that the computed log averages are statistically indistinguishable from the measured log averages, given the uncertainties in the measured logs and in the model computations. To estimate this *probability of fit*, we assume that the combined measurement and model errors are normally distributed for each log and are statistically independent between logs. We can then define a *probability index* P for each log by

$$P = \exp\left(-0.5\left(\left(L - M\right)/E\right)^2\right). \quad (12.7)$$

In Equation 12.7, L is the average of the measured log; M is the modeled log average; and E is the combined measurement and model uncertainty. P is proportional to the probability that L and M are statistically indistinguishable. The *joint probability index*, JPI, for all the logs in the analysis is then the product of the individual indices:

$$JPI = P_1 \cdot P_2 \cdot \cdots \cdot P_N \quad (12.8)$$

Numerically, JPI is equal to 1 if every modeled value (M) is exactly equal to its corresponding log value (L); every deviation from equality moves JPI closer to zero. Overall, JPI is proportional to the probability that *all* the modeled log averages are statistically indistinguishable from their corresponding averaged log measurements.

Updating the statistical earth model. The joint probability index, JPI, is used to weight each realization to construct a final statistical earth model that reflects the effect of imposing the well-log constraints on the *a priori* earth model. One way to do this is to convert JPI into a count: For example, multiply JPI by 10 and truncate to the next lower integer, N. Then count the realization N times in the final earth model. Each realization is thus counted in proportion to its probability of fit. Realizations with JPI < 0.1 receive a count of zero and do not appear in the final earth-model distributions, while a realization that fits perfectly would be counted ten times.

Enough realizations? The Monte Carlo inversion can be stopped when there are enough independent realizations to adequately define the final distributions of the model parameters. The required number of realizations increases with the number of independent model parameters. A large proportion of the realizations have near-zero JPI and are discarded in the construction of the final distributions. Typically it may take tens of thousands of realizations to achieve adequate definition.

Implementation

Implementation of VLSA requires a software application that performs all the operations depicted in Figure 12.1. Table 12.1 gives a functional description of these requirements.

VLSA takes only single interval-average well log values as inputs, in contrast to more conventional techniques that use depth-by-depth log readings. This low density of input data lends itself to implementation in a spreadsheet format, and ExxonMobil uses this approach. Log averages are computed in a separate log analysis application and transferred manually to a data-entry worksheet in the VLSA application.

Example

The following example was introduced in Chapter 2. Here we describe in more detail the VLSA analysis of the data illustrated again in Figure 12.2, where the focus is on a 14.5-m [47.6-ft] shaly interval between two thick channel sands in a deep-water reservoir. If this shaly interval had not been cored, its production potential might never have been recognized. Analysis of ultraviolet core photographs shows 65% net-to-gross, and core plug measurements of porosity and water saturation yield 1.5 m of HPT in this interval. On the other hand, conventional log analysis indicates only 15% net-to-gross and 0.3 m [1 ft] of HPT.

Table 12.1. Functions of a software application implementing VLSA.

Function	Description
Build *a priori* statistical earth model	Provide a straightforward, general mechanism for describing *a priori* distributions of petrophysical properties. (One approach, illustrated in the example below, is to use truncated normal distributions).
Build VLSA analysis model	Provide flexible mechanism to select specific logs to be included in a given analysis model.
Compute interval-average log response models	Incorporate a library of interval-average models for all logs that might be used in low-resolution analysis.
Generate realizations	Incorporate a high-quality random number generator to produce the set of independent random values for each realization.
Compare measured and computed logs	Compute a *probability of fit* like that described by Equation 12.8.
Perform Monte Carlo inversion	Store the results of realizations with positive weight; provide control over number of realizations generated.
Generate final statistical earth model	Calculate summary statistics and generate graphs depicting both final and *a priori* distributions of petrophysical properties including net reservoir, porosity, water saturation, and HPT.

Conventional log analysis of the depicted interval uses a cutoff on the gamma ray (GR) to identify net sand. Porosity is computed from the bulk-density log (RHOB) and water saturation is derived from the deep induction resistivity via a simple Archie analysis.

The VLSA analysis requires two sets of inputs: The interval averages of the well logs plus their associated uncertainties, and the *a priori* statistical earth model of the thin-bedded formation. We discuss these inputs next.

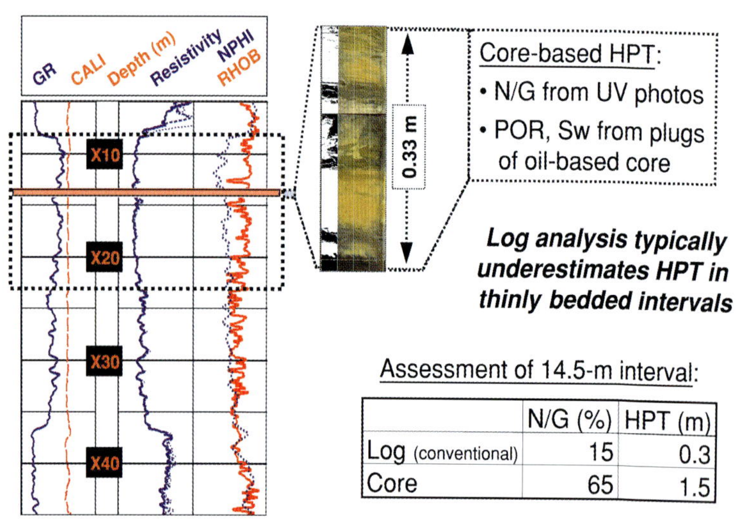

Figure 12.2. Log and core example of thinly bedded pay between two massive sandstones. Note mismatch of HPT from core and conventional log analysis methods for the 14.5-m [47.6-ft] interval outlined on the Figure.

Table 12.2 shows the first set of inputs: The log averages computed across the 14.5-m [47.6-ft] analysis interval and the uncertainties assigned to each average. The uncertainties (labeled "Std Dev") represent the combined measurement uncertainty and model uncertainty for the interval average of each log. The determination of these uncertainties, particularly the model uncertainty, is somewhat subjective. Note that the assigned values are larger than the advertised accuracies for each of the three logs.

The relative uncertainties ("Rel Std Dev") determine the relative weighting of the logs in the Monte Carlo inversion, and this relative weighting has more effect on results than the absolute values of the uncertainties. Table 12.2 shows that the gamma ray log, with its larger relative uncertainty, is weighted significantly less than the bulk density and conductivity.

Table 12.2. Interval-average log inputs for VLSA.

Log	Mean	Std Dev	Unit	Rel Std Dev
Gamma Ray	87	10	GAPI	11.5%
Bulk Density	2.314	0.03	g/cm^3	1.3%
Conductivity	642	15	mmho/m	2.3%

The second set of inputs describes the *a priori* statistical earth model for the formation. One simple and fairly general way to represent the *a priori* distributions for the petrophysical parameters is by means of truncated normal distributions. This representation requires mean, standard deviation, minimum, and maximum values for each parameter. Figure 12.3 illustrates three normal distributions with mean value 0.15, each truncated at minimum and maximum bounds of 0.05 and 0.30, and with varying standard deviations. As the figure shows, a large standard deviation produces an essentially uniform distribution bounded by the minimum and the maximum values.

Table 12.3 shows the complete set of parameters for the *a priori* statistical earth model for our example, where each distribution is represented as a truncated normal. The mean values are either standard log-analysis parameters (e.g., brine conductivity, cementation exponent, etc.), or they are unknowns (e.g., sand and shale fractions) that are obtained in a preliminary step of the VLSA analysis by a least-squares optimization procedure. The standard deviation, minimum, and maximum are assigned to best represent the state of *a priori* knowledge about the range of variability and the uncertainty in each parameter.

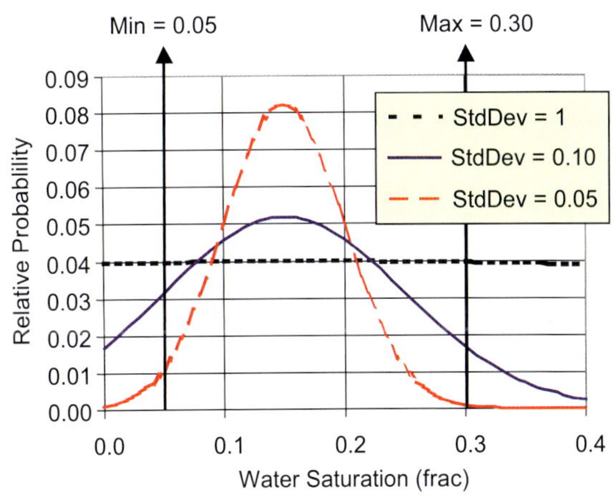

Figure 12.3. Truncated normal distributions.

Table 12.3. *A priori* distributions of petrophysical parameters.

Parameter		Unit	Definition	Mean	Std Dev	Min	Max
Cw		mmho/m	brine conductivity	55555	5000	45555	65555
Angle		degrees	relative hole angle	0	1	0	0
h(m)		meters	interval thickness	14.5	1	14.5	14.5
Vol	Sand	fraction	volume fraction	0.57	1	0.00	1.00
	Shale	fraction	volume fraction	0.43	1	0.00	1.00
PHIT	Sand	fraction	total porosity	0.25	0.02	0.22	0.28
	Shale	fraction	total porosity	0.13	0.02	0.10	0.16
SWT	Sand	fraction	water saturation	0.25	1	0.05	0.45
m	Sand	none	cementation exp.	1.80	0.05	1.70	1.90
n	Sand	none	saturation exp.	1.80	0.05	1.70	1.90
Ct	Shale	mmho/m	shale conductivity	1000	200	900	1100
rhoma	Sand	g/cm^3	matrix density	2.66	1.00	2.66	2.66
	Shale	g/cm^3	matrix density	2.66	1.00	2.66	2.66
rhof	Sand	g/cm^3	fluid density	0.90	0.50	0.90	0.90
	Shale	g/cm^3	fluid density	1.00	0.50	1.00	1.00
GR	Sand	GAPI	GR end-point	30	10	30	30
	Shale	GAPI	GR end-point	150	20	150	150

Where the Min and Max are both set equal to the Mean, the parameter is held constant during inversion. For some parameters, such as the borehole angle and the interval thickness, this assignment reflects an assumption that the parameter is known with sufficient accuracy that it does not affect the uncertainty in the analysis of the reservoir interval.

Alternatively, the uncertainty for some model parameters may be absorbed into the corresponding model uncertainty, as was done in this example with the matrix density, fluid density, and GR end-points. This simplification reduces the number of independent random parameters generated in the Monte Carlo inversion, and hence reduces the computation time.

For this example, sandstone porosity and cementation and saturation exponents are assumed fairly well known from core data. Shale porosity and conductivity are obtained from log readings in thick shales.

Two important parameters in Table 12.3 have almost no *a priori* information imposed on them: the sand and shale volume fractions are uniform distributions between 0 and 1, and the sand water saturation is a uniform distribution between 0.05 and 0.45. The VLSA procedure is thus expected to extract more precise values for these parameters from the log data.

The Monte Carlo inversion procedure for this example required about 28,000 realizations to produce 1000 independent samples, each of which was counted in the final distributions with a weight between 1 and 9 depending on its probability of fit. Figure 12.4 shows these distributions for the major sandstone parameters of volume fraction, porosity, and water saturation, and for the derived HPT solution. In each case the empty histogram bars represent the *a priori* distribution, the red histogram is the final distribution, and the tabulated statistics are for the final distribution. The P5, P10, P50, P90, and P95 statistics are the percentiles of the final distributions.

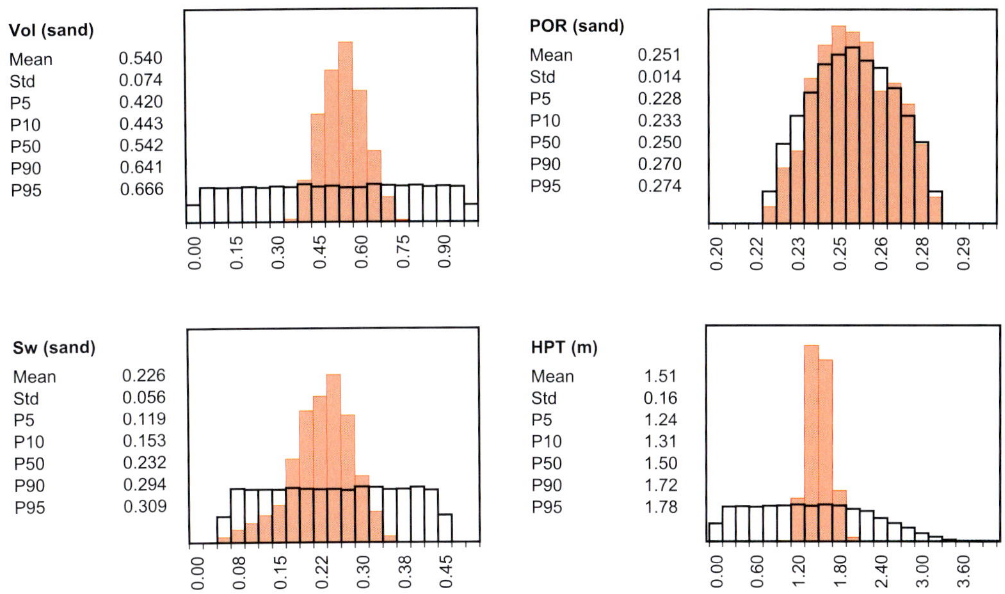

Figure 12.4. ***A priori*** **and final distributions for key model parameters.**

Note that the *a priori* and final distributions of sandstone porosity are essentially the same. With its assigned uncertainty of 0.03 g/cm^3, the bulk density log did not improve on the core-based *a priori* estimate of average sandstone porosity. If a smaller uncertainty had been assigned to the density, the final porosity histogram would be narrower.

On the other hand, the distributions of sand volume fraction and water saturation have been narrowed significantly. The sand fraction has a mean value of 0.54 and a standard deviation of 0.074, or 14% of the mean. Sand water saturation has a mean value of 0.23 and a standard deviation of 0.056, or 24% of the mean. The higher relative uncertainty in water saturation reflects the fact that in a thinly bedded reservoir, the resistivity log is more sensitive to shale fraction than to sand water saturation.

Hydrocarbon pore-thickness (HPT) is derived from sandstone fraction, porosity, and water saturation. In the VLSA solution illustrated in Figure 12.4, HPT has a mean value of 1.51 m [4.95 ft] and a standard deviation of 0.16 m [0.52 ft] — only 11% of the mean.

One might question how it is possible for HPT to have a smaller relative error than some of its constituent factors. The answer is found in an examination of the post-inversion correlations between these factors. The original randomly generated *a priori* samples of sand fraction and sand water sat-

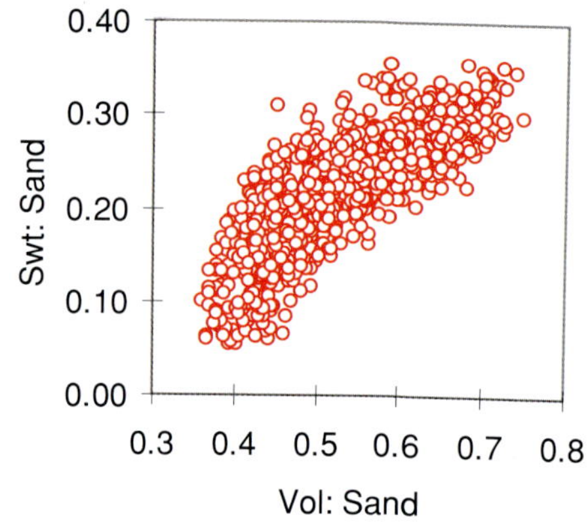

Figure 12.5. Correlation between sand fraction and water saturation in final distributions.

uration are statistically independent. However, as Figure 12.5 shows, the well logs impose a distinct correlation between the sand fraction and the sand water saturation in the final inverted dataset. As sand fraction increases, water saturation must also increase to satisfy the model equations; this correlation narrows the range of computed HPT.

As noted in Chapter 2, the VLSA-derived HPT in this example was verified by a detailed core-based analysis which also yielded 1.5 m [4.9 ft] of hydrocarbons. Conventional log analysis, on the other hand, estimated only 0.3 m [1 ft] for the same interval. Application of a high-resolution approach to this interval is essentially ruled out by the prevalence of beds less than 6 in. [15 cm] thick, as seen in core photographs (Figure 12.2).

Advantages and limitations

ADVANTAGES:

- VLSA can significantly improve the accuracy of HPT evaluations of thinly bedded reservoirs, using only standard low-resolution well logs.
- VLSA explicitly analyzes the uncertainty inherent in the evaluation of thinly bedded reservoirs.
- VLSA can account for the effect of dipping beds and can approximate the effect of finite lateral bed dimensions for induction logs.
- *A priori* information about petrophysical parameters can be incorporated directly into the VLSA analysis, potentially reducing the uncertainty.
- In contrast to high-resolution modeling techniques, VLSA does not require the identification of individual thin beds and does not require painstaking depth-alignment of logs.
- In contrast to earlier laminated-sand analysis methods, VLSA does not rely on low-precision estimates of shale fraction to make error-prone "corrections" to the induction resistivity for the effect of thin shale beds.

LIMITATIONS:

- VLSA only provides interval-average solutions; no depth-by-depth log output is generated.
- VLSA depends on a restrictive set of assumptions about the linearity of log responses. Essentially it is assumed that all the logs can be adequately described by convolution models.
- Invasion and borehole effects are not incorporated.

- The existence of a thinly bedded reservoir must be confirmed independently. Wrong results can be obtained by applying VLSA to a non-productive interval that may appear on conventional logs like a thinly bedded reservoir (for example, a low-quality "dispersed" shaly sand with low hydrocarbon saturation).

Recommendations

WHEN TO APPLY:

- If log responses satisfy the required assumptions, then VLSA may be applied to any siliciclastic reservoir that is confirmed independently to have thin and/or very thin beds.
- In the application of VLSA it is explicitly assumed that borehole and invasion effects are not significant.
- Spikes or horns on induction logs indicate non-linearities in response that cannot be modeled with VLSA.
- If resistivity is measured by a focused-current contact log instead of an induction log, then dipping beds cannot be modeled in VLSA using methods discussed in this chapter.

MODEL SETUP:

- Care should be taken, in computing interval-average log values, to avoid shoulder-bed effects.
- Average conductivity must be computed as the average of depth-by-depth conductivity values, not as the reciprocal of an average resistivity.
- Uncertainty values for log inputs should be related to the accuracy of each log measurement and should be assigned to reflect the desired relative weight to be given to each log.
- If no *a priori* data are available to define the most probable value of a particular petrophysical parameter, then it is modeled as a uniform distribution between reasonable upper and lower bounds.
- In order to minimize the uncertainty in analysis results, all available independent information, such as core data, should be incorporated into the *a priori* parameter distributions.

MODEL EXECUTION:

- To properly define final parameter distributions, enough realizations should be run to generate at least 500 independent samples with non-zero probability of fit.

Other Techniques: Special Logs

NMR

Under appropriate circumstances, Nuclear Magnetic Resonance (NMR) logs provide valuable low-resolution data for the detection and evaluation of thinly bedded reservoirs. The physics of NMR logging is complex, and it is beyond the scope of this publication to explain NMR technology from the beginning. We start here with the fact that the raw NMR signals, recorded during logging, may be processed to obtain an *NMR T_2 distribution* at each logged depth [Prammer, 1994]. T_2 is a measure of the time (in milliseconds) required for a proton in the formation to return to its original energetic state after being excited by an external electromagnetic field. The distribution of T_2 times is influenced by several factors, but under appropriate conditions, the most significant factor is the logged formation's *pore size distribution* [Miller et al., 1990]. Short T_2 times are associated with small pores, while long times are associated with larger pores. A hypothetical T_2 distribution curve is pictured in Figure 12.6.

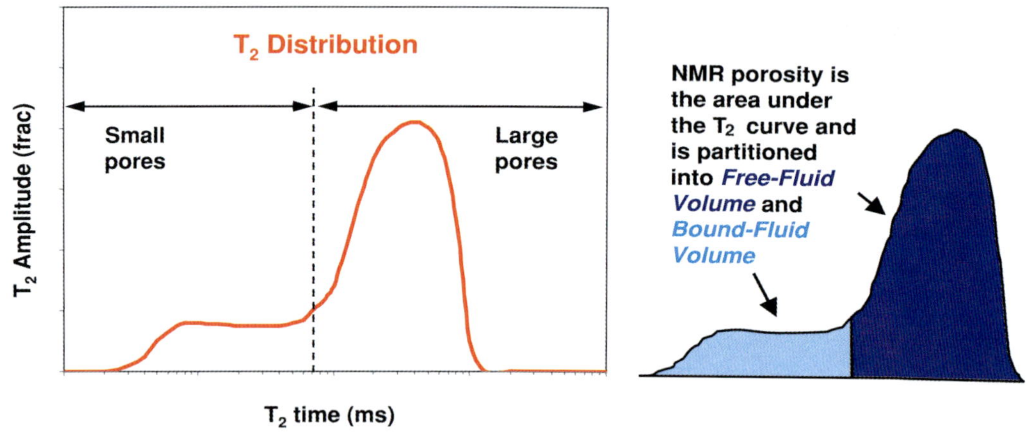

Figure 12.6. T_2 distribution curve showing small and large pores.

NMR-based porosity is the area under the T_2 distribution curve. A cutoff, indicated by the dashed line in Figure 12.6, is applied to the T_2 distribution to determine porosity in small pores and large pores. The former is referred to as bound-fluid volume and the latter is referred to as free-fluid volume.

The first, qualitative application of NMR is to help confirm the presence of thin beds in intervals that appear shaly on conventional logs. Figure 12.7 shows a 55-m [180-ft] logged interval in a deep-water reservoir, with two sections outline by red boxes. These two sections have similar shaly appearance on the gamma ray and resistivity logs. The NMR T_2 distributions, plotted in the right track, show a distinct difference between the two zones. The upper zone has no free fluid (as indicated by the flat distribution curves to the right of the red cutoff line) while the lower zone has significant free-fluid volume.

In Figure 12.7 the upper zone is a low-permeability debrite with no hydrocarbon saturation, while the lower zone is a thinly bedded turbidite with significant sand fraction and hydrocarbons in the interbedded sands, despite its low resistivity. The gamma ray and resistivity logs show only a rather weak and ambiguous distinction between these two intervals, while the NMR T_2 distribution differentiates them clearly.

The second, quantitative application of NMR is to obtain a direct indication of HPT for a thinly bedded interval. In a light oil-bearing siliciclastic reservoir at irreducible water saturation, the free fluid volume equals the moveable hydrocarbon volume, or $\phi(1\text{-}S_w)$. Thus the hydrocarbon pore-thickness may be estimated as the accumulated value of log sample spacing times the free-fluid volume readings across the reservoir interval:

$$HPT = h\sum_i FF_i \tag{12.9}$$

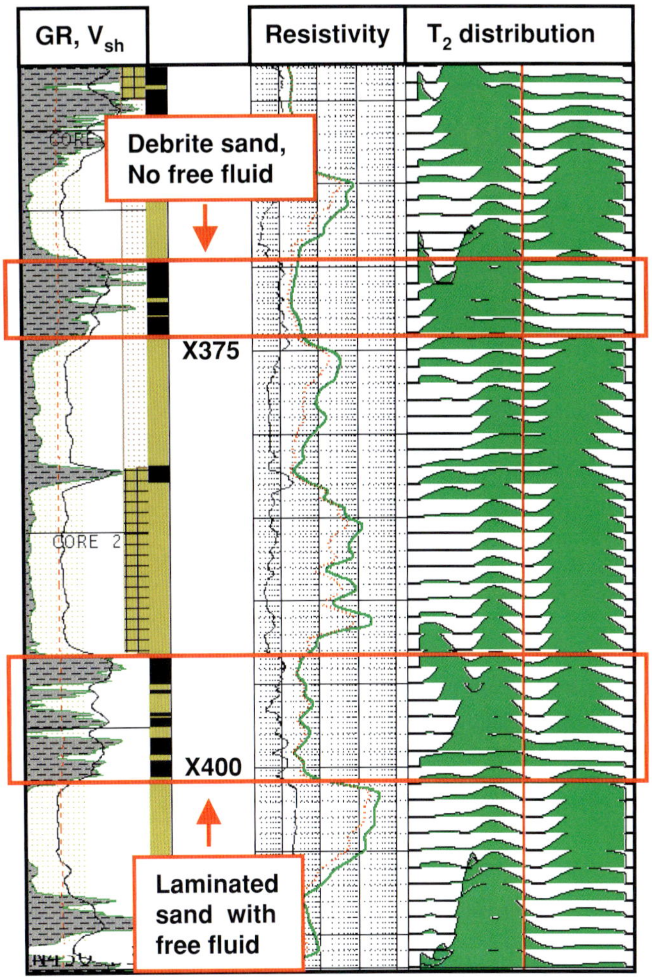

Figure 12.7. NMR T_2 distribution distinguishes between low-permeability muddy debris flow and laminated sand.

In Equation 12.9, h is the spacing between log samples and the FF_i are the sample values of NMR free-fluid fraction at each depth in the reservoir interval. Figure 12.8 shows an example where NMR confirmed the existence of thin beds and an increase in HPT compared to conventional cutoff-based resistivity log analysis.

In Track 1 of Figure 12.8, the gamma ray (GR, thin black curve) and the GR-based shale volume (light grey shading) indicate significant shaliness in the red boxed interval from X240 to X255 meters. The deep laterolog resistivity (thin black curve to the right in Track 2) is suppressed across this whole interval relative to the thick sands below. In addition, the density and neutron porosities indicate some shaliness over most of the interval. (Note also the apparent light-hydrocarbon cross-over of these two curves in the thicker sands).

Despite the conventional indications of shaliness, the NMR T_2 distributions plotted in Track 3 show significant free-fluid volume from about X241 to X255, as indicated by the green shading to the right of the red vertical cutoff line. The free fluid volume is also plotted as a continuous log with green shading in Track 5. The density porosity is plotted there with gray shading. The NMR free fluid volume provides strong evidence for the presence of hydrocarbon-saturated thin beds in the shaly interval from X240 to X255 m, even though the NMR log clearly cannot resolve individual thin beds.

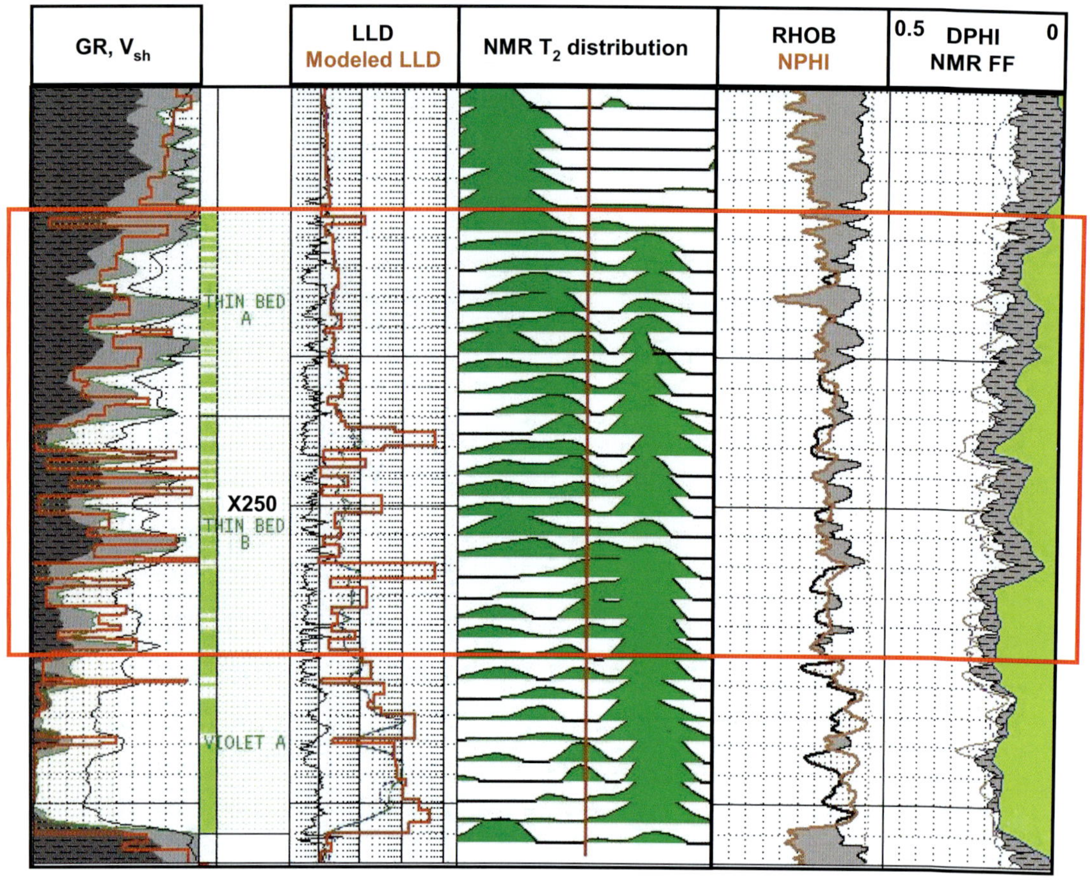

Figure 12.8. GR-based shale volume (light grey shading in Track 1) indicates shaliness from X240 to X255 m (boxed interval); NMR T_2 distribution shows abundant free-fluid volume across the interval.

The 21-m [69-ft] reservoir interval in Figure 12.8 was analyzed by four methods which are summarized in Table 12.4.

Table 12.4. Results of four methods of HPT analysis.

Analysis method	HPT (m)
Conventional, with GR cutoff to identify net sand	2.0
Resisitivity modeling applied to LLD	2.3
VLSA	2.7
HPT from NMR free-fluid volume (Equation 12.9)	2.7

The modeled resistivity, plotted in red in Track 2 of Figure 12.8, improved the resistivity values in sand beds thicker than about half a meter. Nonetheless, the modeled resistivity is still suppressed across the interval from X240 to X247, where the NMR shows significant free-fluid volume and where beds are evidently too thin for detailed modeling.

Note the agreement between VLSA and the NMR free-fluid volume, which are derived from completely separate and independent measurements. Both methods produced an additional 17% increase in HPT over that obtained via resistivity modeling.

Multi-component induction

As discussed in earlier chapters, a primary reason for the difficulty in analyzing thinly bedded reservoirs with conventional logs is that standard induction and focused-current logs measure only one component of the anisotropic formation conductivity. In the typical case (borehole perpendicular to bedding planes), the measured component is oriented parallel to the bedding planes and depends much more strongly on the conductive interbedded shales than on the resistive hydrocarbon sands.

The multi-component induction log offers the possibility of measuring the conductivity component perpendicular to bedding planes, which is very sensitive to the resistive sand beds. Here we present a synthetic VLSA example to illustrate the reduction in uncertainty that may be possible when this measurement is available. The synthetic dataset was introduced in Chapter 2 for a VLSA example using only the conventional induction-log data. Figure 12.9 shows the synthetic log data including the simulated "vertical resistivity" in Track 2.

Table 12.5 shows the interval-average log data and the assigned uncertainties for the 20-ft [6-m] thinly bedded interval in Figure 12.9. In the absence of specific information about the accuracy of the vertical (transverse) component of conductivity measured by a multicomponent induction tool, the same absolute uncertainty was assigned to this component as to the conventional horizontal (parallel) component. Note that the resulting relative uncertainty is 16%, which is quite high.

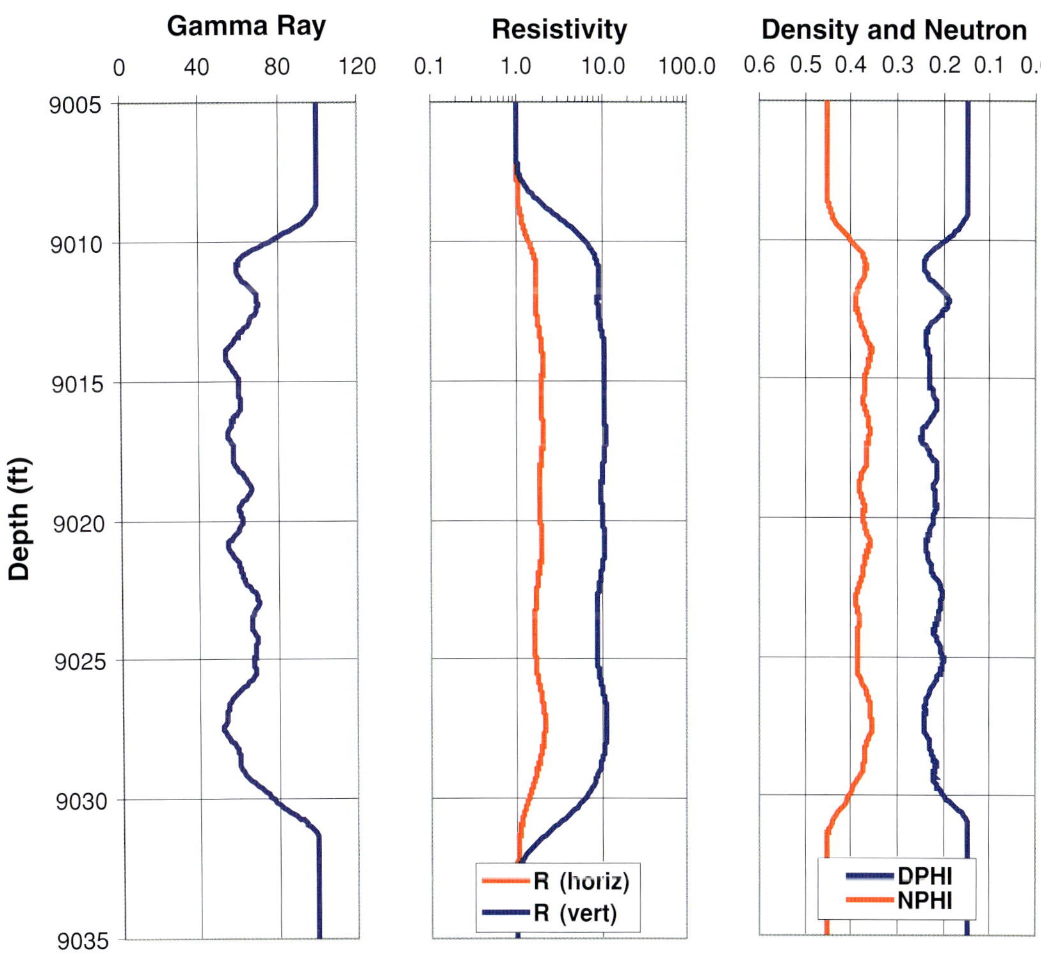

Figure 12.9. Synthetic log example with horizontal (red) and vertical (blue) resistivity components.

Table 12.5. Interval-average log inputs for VLSA

Log	Mean	Std Dev	Unit	Rel Std Dev
Gamma Ray	60	10	GAPI	17%
Bulk Density	2.238	0.02	g/cm^3	0.9%
Horiz. Conductivity	525	15	mmho/m	2.9%
Vert. Conductivity	95	15	mmho/m	16%

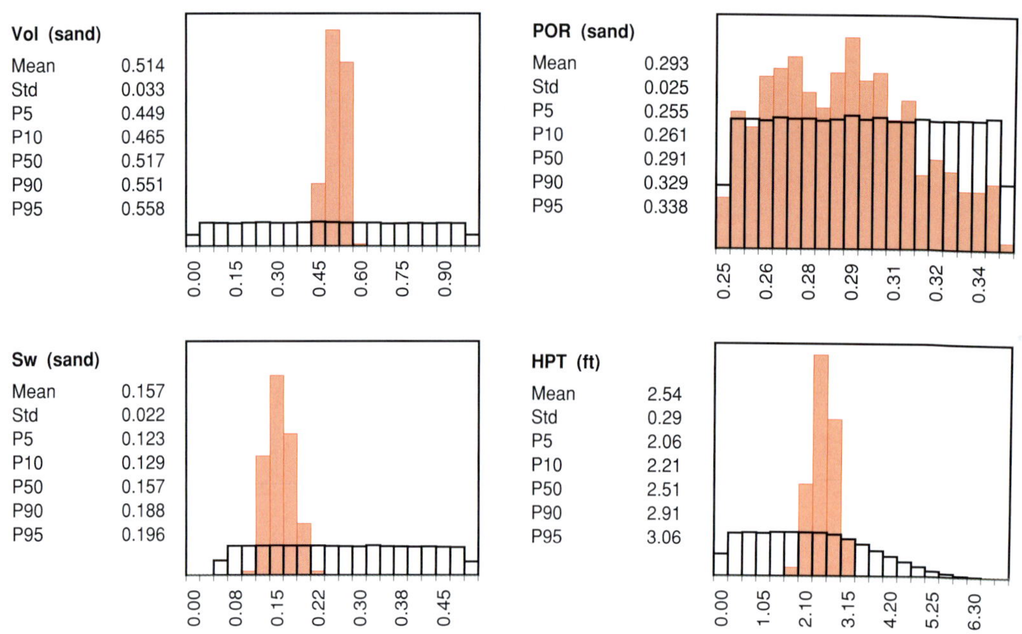

Figure 12.10. A priori and final distributions for VLSA analysis incorporating transverse component of formation conductivity.

With the addition of the constraint imposed by the vertical conductivity measurement, the VLSA Monte Carlo inversion produced 340 independent samples in 100,000 trials. The resulting final distributions for the main reservoir parameters are shown in Figure 12.10, which should be compared with Figure 2.5.

Figure 12.10 may be compared with Figure 2.5, which shows the same final distributions for a VLSA analysis of the synthetic data utilizing only the conventional parallel component of conductivity along with the gamma ray and the bulk density. Table 12.6 compares the mean values and the relative uncertainties of the sand fraction, porosity, water saturation, and HPT for the two cases.

Note that the addition of the transverse conductivity has yielded significant improvements in the absolute accuracy of the VLSA-derived mean sand water saturation and sand fraction, and a smaller improvement in the interval HPT. In addition the VLSA-estimated relative uncertainties in these parameters were reduced substantially, despite the high relative uncertainty assigned to the transverse conductivity measurement. Thus, it appears that the transverse conductivity may contribute significantly to the evaluation of thinly bedded reservoirs, even if its measurement accuracy is less than that of the conventional parallel conductivity.

Table 12.6. Mean values and relative uncertainties for VLSA example with and without transverse conductivity.

Parameter	True value	Parallel conductivity only		Parallel and transverse conductivity	
		VLSA mean	Relative uncertainty	VLSA mean	Relative uncertainty
Sand fraction	0.50	0.58	12%	0.51	6%
Sand porosity	0.30	0.29	7.9%	0.29	8.5%
Sand S_w	0.15	0.28	39%	0.16	14%
Interval HPT	2.50	2.40	13%	2.54	11%

Advantages and limitations

NMR: ADVANTAGES

- Even in the absence of high-resolution data, the NMR log can provide strong evidence for the existence of hydrocarbon-bearing thin beds in intervals that appear shaly on conventional logs.
- Under appropriate conditions, NMR provides a direct estimate of HPT that is completely separate and independent from conventional resistivity, porosity, and lithology logs.

NMR: LIMITATIONS

- The NMR T_2 distribution may be affected by numerous factors other than pore size. These factors include fluid type, fluid viscosity, and lithology. Therefore the application of NMR for HPT determination in a thinly bedded reservoir requires independent knowledge of these factors.

MULTI-COMPONENT INDUCTION: ADVANTAGES

- Measurement of the vertical, or transverse, component of resistivity can provide strong evidence for the presence of hydrocarbon-bearing thin beds in intervals that appear shaly on conventional logs.
- The transverse-resistivity log can significantly reduce the uncertainty in the low-resolution evaluation of a thinly bedded reservoir.

MULTI-COMPONENT INDUCTION: LIMITATIONS

At the time of writing (late 2004) the multi-component induction tools are not yet in widespread use and many aspects of their practical functioning have not been widely understood and tested independent of the service companies who have developed these tools. The limitations mentioned here reflect this relative newness and shortage of operator experience.

- Multi-component resistivity logs are not widely available.
- The accuracy of the transverse resistivity measurement, and its dependence on environmental factors, is not well known.

Summary

Low-resolution evaluation refers to any method of evaluating thin-bedded reservoirs that does not resolve the properties of each individual thin bed. We recognize three categories of low-resolution evaluation specific to thin-bedded reservoirs: First, variants of the conventional laminated sand analysis, whose shortcomings were documented in Chapter 1; second, a new method called Volumetric Laminated Sand Analysis (VLSA), which is detailed in this chapter; and third, application of two specific low-resolution logs that have particular utility for evaluating thin beds, also discussed in this chapter.

The *Volumetric Laminated Sand Analysis* (*VLSA*) procedure begins with an *a priori* statistical earth model that describes the potential range of petrophysical properties for each bed type, before the introduction of well-log information. Interval-average forward models are used to predict interval-average well-log response for each instance of the statistical earth model. Monte Carlo inversion of these models is employed to update the statistical earth model, generating final distributions of petrophysical properties that are consistent with the well-log data. These final distributions yield a solution for reservoir HPT plus a description of the uncertainty in the solution.

VLSA can significantly improve the accuracy of HPT evaluations of thinly bedded reservoirs, using only standard low-resolution well logs. Further, VLSA explicitly analyzes the uncertainty inherent in thin-bed evaluation. The main limitations of VLSA are that it generates only an interval-average solution with no depth-by-depth log output; and that the method depends on a restrictive set of assumptions about the linearity of log responses.

The low-resolution Nuclear Magnetic Resonance (NMR) log can provide strong evidence for the existence of hydrocarbon-bearing thin beds in intervals that appear shaly on conventional logs. Under appropriate conditions, NMR also provides a direct HPT estimate that is independent from the conventional log suite. However, the NMR log measurement is complex and is subject to numerous disturbing environmental factors.

The low-resolution multi-component induction log can also provide strong evidence for the presence of hydrocarbon-bearing thin beds in intervals that appear shaly on conventional logs. In principle, this tool can significantly reduce the uncertainty in the low-resolution evaluation of a thinly bedded reservoir. At the time of writing, however, the multi-component induction tools are not yet widely available. Their field accuracy and reliability have not yet been thoroughly evaluated independent of the service companies who are developing the tools.

Photo by Rob Wellner

**Tide-influenced Delta,
Cretaceous (Cenomanian),
Frewins Castle Sandstone,
Belle Frourche Member,
Frontier Formation:
Tisdale Mountain Anticline, Wyoming, U.S.A.**

13 Summary and Recommendations

Overview

Introduction

This chapter presents a brief summary of the methods that have been described in earlier chapters for evaluating thinly bedded siliciclastic reservoirs. The summary includes an outline of a recommended procedure, illustrated by a flowchart.

Contents

DOI: 10.1306/1157796A13220

Summary of Methods

Failure of conventional techniques

The direct application of Archie or shaly sand models (e.g., Waxman-Smits, Dual-water) to thinly bedded intervals generally results in underestimation of hydrocarbon pore-thickness (HPT), even when no cutoffs are applied. Shaly sand models, such as the Waxman-Smits and Dual-water models, are designed to address the effects of dispersed clay and, in principle, are not applicable to laminated sands.

Conventional laminated-sand techniques rely on solving the parallel-conductivity equation for sandstone resistivity. These techniques address the thin-bed problem directly, but they are prone to significant error because the resistivity log in a thinly bedded siliciclastic reservoir is much more sensitive to the shale fraction than to the sandstone resistivity.

Two thin-bed approaches

There are two general log analysis approaches for thinly bedded formations:

1) Identify bed boundaries using *high-resolution* data and attempt to resolve true log values in each thin bed through log forward-modeling and inversion techniques. These methods are described in Chapter 11.

2) Analyze the effects of thin beds on standard-resolution or interval-average log responses, and compensate for these effects without resolving the individual beds (the *low-resolution* approach). These methods are described in Chapter 12.

1) High-resolution methods

The application of resistivity forward modeling and inversion is recommended when bed thicknesses exceed 1 or 2 ft [.3 or .6 m], or where the deep resistivity log is affected by invasion, or where beds are dipping more than 30° with respect to the borehole.

When bed thickness is 2–4 ft [.6–1.2 m] or less, thin-bed effects on porosity logs can have a significant impact on HPT estimates [Chapter 1]. Thus, if high-resolution modeling is attempted for beds 2 ft [.6 m] thick or less, it may be necessary to model porosity tool response as well as resistivity tool response.

ADVANTAGES OF THE HIGH-RESOLUTION APPROACH:

- It provides a high-resolution log result and display showing the distribution of thin beds and hydrocarbons.
- It can employ the most detailed and accurate logging tool models available.

LIMITATIONS OF THE HIGH-RESOLUTION APPROACH:

- A high-resolution log is required to identify the boundaries of each thin sand and shale bed.
- The modeling and inversion procedure does not produce a unique result. Therefore there is significant uncertainty in the solution, especially when bed thicknesses are less than the blind frequency of the resistivity log.

- The manual forward-modeling and inversion procedure can be very labor intensive, especially with many thin beds.

The procedure becomes impractical when bed thickness approaches the *very thin* range (less than a few inches thick).

2) Low-resolution methods

Volumetric Laminated Sand Analysis (VLSA) is our primary low-resolution technique utilizing conventional well logs and interval-average response models. Application of a technique like VLSA is recommended in reservoirs where many beds are less than 1 or 2 ft [.3 or .6 m] thick.

ADVANTAGES OF VLSA:

- In contrast to high-resolution modeling techniques, VLSA does not require the identification of individual thin beds and does not require painstaking depth-alignment of logs.
- VLSA explicitly analyzes the uncertainty inherent in the evaluation of thinly bedded reservoirs.
- The application of VLSA is relatively simple and quick.

LIMITATIONS OF VLSA:

- VLSA only provides interval-average solutions; no depth-by-depth log output is generated.
- VLSA depends on a restrictive set of assumptions about the linearity of log responses.
- The existence of a thinly bedded reservoir must be confirmed independently; otherwise misleading results may be obtained.

OTHER METHODS: NMR AND MULTI-COMPONENT INDUCTION

Under appropriate conditions, Nuclear Magnetic Resonance (NMR) logs can provide independent evidence for the presence of thin beds and an independent estimate of hydrocarbon pore-thickness. Thus the NMR log is a useful measurement to help resolve uncertainty when the presence of thin beds is anticipated or suspected in a siliciclastic light-oil reservoir.

The relatively new multi-component induction tools can, in principle, provide a measure of the transverse (series) resistivity of a thinly bedded formation. This series resistivity is much more sensitive to the thin, resistive sandstone beds than the parallel resistivity, the quantity measured by conventional induction and focused-current resistivity logs. The multi-component resistivity tools may prove extremely valuable, both for identifying thinly bedded reservoirs and for reducing the uncertainty inherent in their evaluation.

Recommendations

Procedure

In the foregoing chapters we have described a comprehensive set of techniques and procedures for evaluating thinly bedded siliciclastic reservoirs. Here we present a brief, high-level summary of these procedures in the form of a series of steps, which are further summarized as a flowchart in Figure 13.1. Associated details are discussed in the referenced chapters.

1) PLAN LOGGING AND CORING PROGRAM [Chapters 4, 5]. Where the potential exists for encountering significant thinly bedded pay [Chapter 3], plan an appropriate logging program. This should include an electrical borehole image log [Chapter 9] or other high-resolution measurement to identify bed boundaries. Consider running NMR and/or multicomponent induction tools to help identify and evaluate thinly bedded intervals [Chapters 4, 5]. Consider coring one or more wells to provide ground truth for the identification and quantification of both reservoir and non-reservoir bed types [Chapters 7, 8].

2) DETERMINE IF THIN BEDS ARE PRESENT AND RESOLVED BY AVAILABLE DATA [Chapters 4, 8, 9]. Core and core photographs provide the most direct identification and resolution of thin beds. Electrical borehole images can often resolve beds down to 1 or 2 in. [2.5 or 5 cm]. Conventional logs cannot resolve beds thinner than 1 or 2 ft [.3 or .6 m], but they may provide strong evidence for the existence of thin or very thin beds, especially if combined with NMR or multi-component induction data. If there is strong evidence for the presence of reservoir intervals where many beds cannot be resolved by available data, then the application of a low-resolution evaluation technique is indicated [Chapter 12].

3) BEDS RESOLVED; MANY BEDS THINNER THAN 1 OR 2 FT [.3 OR .6 M]. In intervals where the dominant bed thickness ranges from several inches up to about 2 ft [.6 m], high-resolution evaluation methods may be applied but their reliability is marginal [Chapter 11]. A low-resolution technique should be applied for confirmation of results and evaluation of uncertainty [Chapter 12].

4) BEDS RESOLVED; MANY BEDS THINNER THAN 4 FT [1.2 M]. Where the dominant bed thickness is roughly between 1 and 4 ft [.3 and 1.2 m], resistivity modeling may be applied with increased confidence [Chapter 11]. In this range, log convolution modeling should also be considered to improve the resolution of lithology and porosity logs [Chapter 11]. A low-resistivity evaluation can increase confidence in HPT results [Chapter 12].

5) BEDS RESOLVED; MANY BEDS THINNER THAN 20 FT [6 M]. If bed thicknesses range from about 4 to 20 ft [1.2 to 6 m], resistivity modeling can be applied with confidence and can significantly improve the estimation of HPT through reducing shoulder-bed effects [Chapter 11]. In this range the effects of shoulder beds on lithology and porosity logs are less significant.

6) BEDS THICKER THAN 20 FT [6 M]. Where reservoir beds are mostly 20 ft [6 m] thick or more, standard log analysis techniques are generally adequate. However, if the resistivity log is affected by deep invasion or borehole effects, or if the relative dip between the borehole and bedding planes exceeds 30°, then resistivity modeling should be employed to compensate for these effects [Chapter 11].

Flowchart

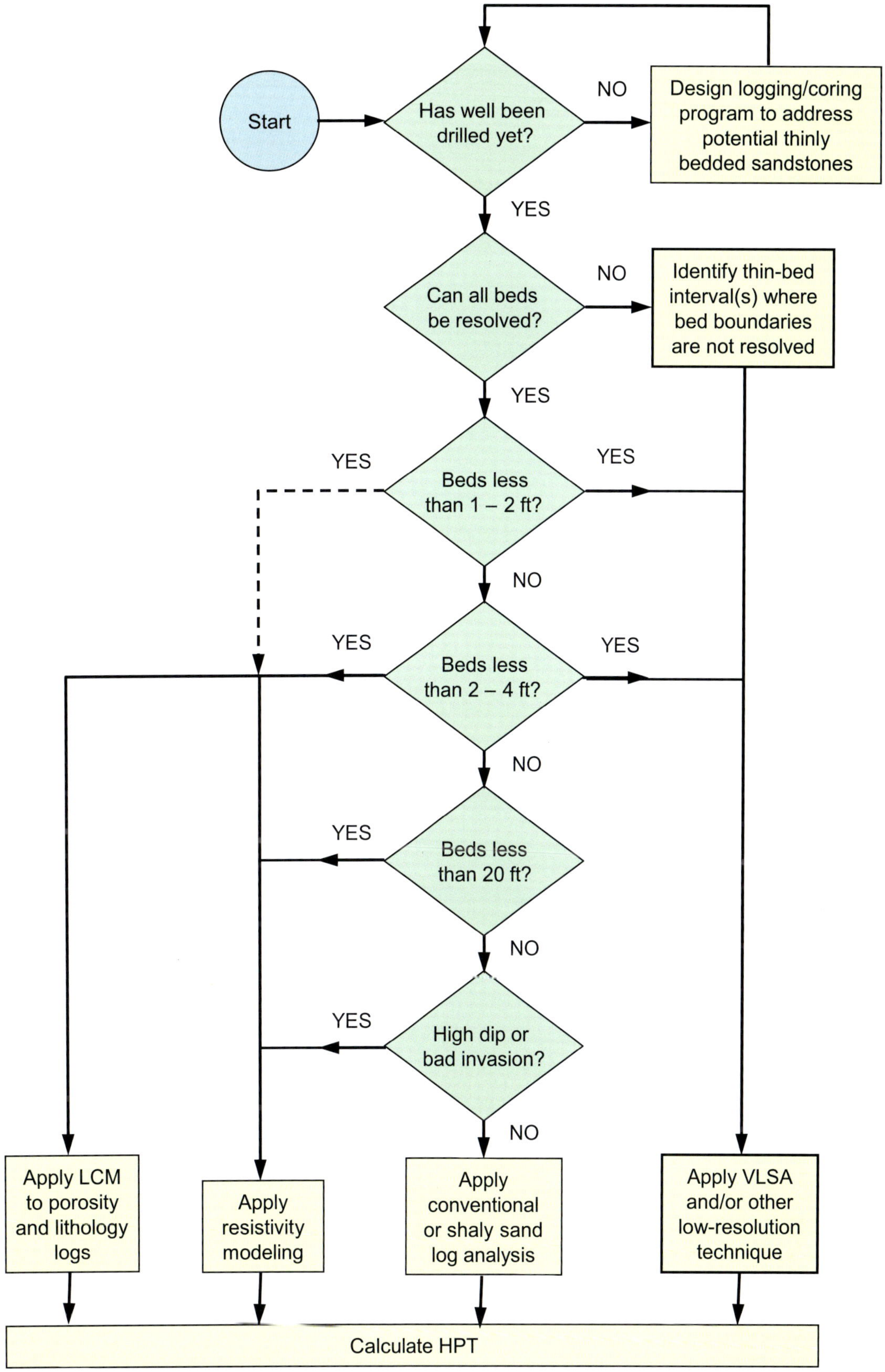

Figure 13.1. Flowchart for evaluation of thinly bedded siliclastic reservoirs.

Bibliography

Adams, J. T., L. T. Bourke, and S. G. Buck, 1990, Integrating Formation Microscanner Images and Cores with Case Studies: Oilfield Review, January, p. 52–65.

Ahmed, U., S. R. Crary, and G. R. Coates, 1989, Permeability Estimation: The Various Sources and their Interrelationship: Society of Petroleum Engineers, 64th Annual Technical Conference and Exhibition, SPE Paper no. 19604, p. 649–692

Akkurt, R., M. A. Moore, and J. J. Freeman, 1997, Impact of NMR in the Development of a Deepwater Turbidite Field, *in* Transactions of the SPWLA 38th Annual Logging Symposium, June 15–18, Houston, Texas: Society of Professional Well Log Analysts, Paper SS, 14 p.

Allen, D. F., 1984, Laminated Sand Analysis, *in* Transactions of the SPWLA 25th Annual Logging Symposium: Society of Professional Well Log Analysts, Paper XX, 20 p.

Anderson, B. and S. Gianzero, 1983, Induction Sonde Response in Stratified Media: The Log Analyst, v. 24, no. 1, p. 25–31.

Anderson, B. I., 1986, The Analysis of Some Unsolved Induction Interpretation Problems Using Computer Modeling: The Log Analyst, v. 27, no. 5, p. 60–73.

Anderson, B., K. A. Safinya, and T. Habashy, 1988, Effect of Dipping Beds on the Response of Induction Tools: SPE Formation Evaluation, March 1988, p. 29–36.

Anderson, B., and T. D. Barber, 1988, Strange Induction Logs—A Catalog of Environmental Effects: The Log Analyst, v. 29, no. 4, July–August, p. 229–243.

Anderson, B. I., T. D. Barber, J. Singer, and T. Broussard, 1989, Elmod—Putting Electromagnetic Modeling to Work to Improve Resistivity Log Interpretation, *in* Transactions of the SPWLA 30th Annual Logging Symposium: Society of Professional Well Log Analysts, Paper M, 20 p.

Anderson, B., S. Bonner, M. G. Lüling, and R. Rosthal, 1990, Response of 2-Mhz LWD Resistivity and Wireline Induction Tools in Dipping Beds and Laminated Formations *in* Transactions of the SPWLA 31st Annual Logging Symposium, Lafayette, Louisiana: Society of Professional Well Log Analysts, Paper A, p. 461–475.

Anderson, B., T. D. Barber, and M. G. Lüling, 1995, The Response of Induction Tools to Dipping, Anisotropic Formation *in* Transactions of the SPWLA 36th Annual Logging Symposium: Society of Professional Well Log Analysts, Paper D, 12 p.

Anderson, B., T. D. Barber, D. Druskin, P. Lee, and V. E. Dussan, 1996, The Response of Multiarray Induction Tools in Highly Dipping Formation with Invasion and in Arbitrary 3-D Geometries, *in* Transactions of the SPWLA 37th Annual Logging Symposium, June: Society of Professional Well Log Analysts, Paper A, p. 327–344.

Anderson, B. I., T. D. Barber, and S. C. Gianzero, 1998, The Effect of Cross-bedding Anisotropy on Induction Tool Response, *in* Transactions of the SPWLA 39th Annual Logging Symposium, May: Society of Professional Well Log Analysts, Paper B, p. 137–149.

Archie, G. E., 1942: The Electrical Resistivity Log as an Aid in Determining Some Reservoir Characteristics: Transactions of the American Institute of Mining, Metallurgical, and Petroleum Engineering, 146, p. 54–62.

Barber, T. D., 1988, Induction Vertical Resolution Enhancement-physics and Limitations, *in* Transactions of the SPWLA 29th Annual Logging Symposium: Society of Professional Well Log Analysts, Paper O, 18 p.

Basan, P., J. R. Hook, K. Hughes, J. Rathmell, and D. C. Thomas, 1988, Measuring Porosity, Saturation and Permeability from Cores: An Appreciation of the Difficulties: Schlumberger Technical Review, v. 36, no. 4, p. 22–36.

Bateman, R., 1990, Thin-bed Analysis with Conventional Log Suites, in Transactions of the SPWLA 31st Annual Logging Symposium: Society of Professional Well Log Analysts, Paper II, 24 p.

Bates, R. L., and J. A. Jackson, 1984, Dictionary of Geological Terms, 3rd Edition: American Geological Institute, New York, Anchor Books, Doubleday, 571 p.

Beck, G. F., 1995, Forward Modeling of Deep Resistivity Response in Norphlet Wells, *in* Transactions of the SPWLA 36th Annual Logging Symposium: Society of Professional Well Log Analysts, Paper HHH, 12 p.

Bergslien, D., O. M. Burtz, F. M. Haynes, and M. S. Munkholm, 2000, Applications of Resistivity Modeling in Reservoir Development: Examples from Balder Field, Norwegian North Sea: Petrophysics, v. 41, no. 4, p. 281–293.

Bourke, L. T., 1989, Recognizing Artifact Images of the Formation Microscanner, *in* Transactions of the SPWLA 13th Annual Logging Symposium, Denver: Society of Professional Well Log Analysts, Paper WW, 25 p.

Boyd, A., B. Davis, C. Flaum, J. Klein, R. M. Sneider, A. Sibbit, and J. Singer, 1995, The Lowdown on Low-resistivity Pay: Oilfield Review, Autumn 1995, p. 4–18.

Boyeldieu, C. and P. Jefferys, 1988, Formation Microscanner: New Developments, *in* Transactions of the SPWLA 11th European Formation Evaluation Symposium, Norway: Society of Professional Well Log Analysts, Paper G, 16 p.

Branisa, F., 1974, Filtering of Well-log Curves: Geophysics, v. 39, no. 4, p. 545–549.

Buscher, M., 1989, Formation Microscanner: A Direct Comparison of 2-Pad vs. 4-Pad Tool, *in* Transactions of the SPWLA 12th French Section (SAID) International Formation Evaluation Symposium, Paris: Society of Professional Well Log Analysts, Paper O, 10 p.

Campbell, C. V., 1967, Lamina, Lamina Set, Bed, and Bedset: Sedimentology, v. 8, p. 7–26.

Chemali, R., S. Gianzero, R. Stickland, and S. M. Tijani, 1983, The Shoulder Bed Effect on the Dual Laterolog and its Variation with the Resistivity of the Borehole Fluid *in* Transactions of the SPWLA 24th Annual Logging Symposium: Society of Professional Well Log Analysts, Paper UU, 25 p.

Chemali, R., S. Gianzero, and S. M. Su, 1987, The Effect of Shale Anisotropy on Focused Resistivity Devices, *in* Transactions of the SPWLA 28th Annual Symposium, London: Society of Professional Well Log Analysts, Paper H, 23 p.

Craft, B. C., and M. F. Hawkins, 1959, Applied Petroleum Reservoir Engineering: Englewood Cliffs, New Jersey, Prentice-Hall, Inc., 284 p.

Czubek, J. A., 1973, New Theory, Possibilities, and Practice in Digital Interpretation of Gamma-ray Logs, *in* Transactions of the SPWLA 14th Annual Logging Symposium: Society of Professional Well Log Analysts, Paper W, 22 p.

Czubek, J. A., 1986, Quantitative Interpretation of Gamma-ray Logs in Presence of Random Noise, *in* Transactions of the SPWLA 27th Annual Logging Symposium: Society of Professional Well Log Analysts, Paper KKK, 25 p.

Dahlberg, K. E., and D. E. Fitz, 1988, Comparing Log-derived and Core-derived Porosity and Mineralogy in Thinly Bedded Reservoirs: An Integrated Approach, *in* Transactions of the SPWLA 29th Annual Logging Symposium: Society of Professional Well Log Analysts, Paper W, 18 p.

Dahlberg, K. E., 2002, Volumetric Laminated Sand Analysis: U. S. Patent Application no. 20020133323, http://www.uspto.gov/patft (accessed November 21, 2005).

Draper, N. R., and H. Smith, 1982, Applied Regression Analysis: Hoboken, New Jersey, John Wiley & Sons, 709 p.

Doll, H., 1949, Introduction to Induction Logging and Application to Logging of Wells Drilled with Oil Based Mud: Journal of Petroleum Technology, v. 1, no. 6, p. 148–162.

Ekstrom, M. P., C. Dahan, M. Y. Chen, P. Lloyd, and D. Rossi, 1987, Formation Imaging with Microelectrical Scanning Arrays: The Log Analyst, v. 28, no. 3, p. 294–306.

Eubanks, D. L., and M. Oraby, 1997, Characterization of Low-contrast/Low-resistivity Reservoirs with Modern Imaging Tools: Example from the Gulf of Mexico: 1997 AAPG Annual Convention, April 6–9, 1997, Dallas, p. A33.

Fam, M. Y., R. Chemali, D. Seiler, M. Haughland, and W. F. Stewart, 1995, Applying Electrical Micro-Imaging Logs to Reservoir Characterization: Annual Society of Petroleum Engineers Technical Conference and Exhibition Transactions, Dallas, SPE Paper no. 30608.

Flaum, C., J. E. Galford, and A. Hastings, 1989, Enhanced Vertical Resolution Processing of Dual Detector Gamma-gamma Density Logs: The Log Analyst, v. 30, no. 3, p. 139–149.

Frodensen, E. W., and T. Lawrence, 1997, Low-resistivity, Low-contrast Pay: Some Examples from Offshore North West Java, Ardjuna Basin, Indonesia: 1997 AAPG Annual Convention, April 6–9, 1997, Dallas, p. A38.

Fylling, A., 1991, Estimating Conductivity and Saturations in the Sands of Thinly Laminated, Dipping Sand/Shale Sequences, *in* Transactions of the SPWLA 32nd Annual Logging Symposium: Society of Professional Well Log Analysts, Paper O, 25 p.

Galford, J. E., C. Flaum, W. A. Gilchrist, and S. W. Duckett, 1989, Enhanced Resolution Processing of Compensated Neutron Logs: SPE Formation Evaluation, June, p. 131–137.

Gardner, J. S. and J. L. Dumanoir, 1980, Litho-density Log Interpretation, *in* Transactions of the SPWLA 21st Annual Logging Symposium: Society of Professional Well Log Analysts, Paper N, 12 p.

Gianzero, S., and B. Anderson, 1982, A New Look at Skin Effect: The Log Analyst, v. 29, no. 1, p. 93–104.

Gianzero, S., Y. Lin, R. Chemali, and W. Dossey, 1985, The Effect of Sonde Eccentering on Resistivity Tools: An Exact Theoretical Model, *in* Transactions of the SPWLA 26th Annual Logging Symposium: Society of Professional Well Log Analysts, Paper GG, 29 p.

Gianzero, S., D. Kennedy, L. Gao, and L. Sanmartin, 2002, The Response of a Triaxial Induction Sonde in a Biaxial Anisotropic Medium: Petrophysics, v. 43, no. 3, p. 172–184.

Glowinski, R., G. F. Rouault, and J. R. Tabanou, 1988, SP Deconvolution and Quantitative Interpretation in Shaly Sands, *in* Transactions of the SPWLA 28th Annual Logging Symposium, London, England: Society of Professional Well Log Analysts, Paper SS, 25 p.

Graciet, S., and L. C. Shen, 1997, Simulation of Induction and MWD Resistivity Tools in Anisotropic Dipping Beds, *in* Transactions of the SPWLA 38th Annual Logging Symposium: Society of Professional Well Log Analysts, Paper M, 14 p.

Hagiwara, T., 1995, Anisotropic Shale and Induction Log Shoulder Bed Corrections for Deviated Boreholes, *in* Transactions of the SPWLA 36th Annual Logging Symposium: Society of Professional Well Log Analysts, Paper Z, 12 p.

Hornby, B. E., L. M. Schwartz, and J. A. Hudson, 1993, Effective Medium Modeling of the Electrical and Elastic Properties of Anisotropic Porous Media, *in* SEG Annual Meeting Transactions: Society of Exploration Geophysicists, p. 786–791.

Howell, E. P. and T. E. Fisher, 1982, Induction Log Deconvolution for Deviated Boreholes, *in* Transactions of the SPWLA 23rd Annual Logging Symposium, June: Society of Professional Well Log Analysts, Paper H, 14 p.

Isaaks, E. H., and R. M. Srivastava, 1989, An Introduction to Applied Geostatistics: Oxford, U.K., Oxford University Press, 592 p.

Jacobson, L. A., D. F. Wyatt, L. L. Gadeken, and G. A. Merchant, 1990, Resolution Enhancement of Nuclear Measurements through Deconvolution, *in* Transactions of the SPWLA 31st Annual Logging Symposium: Society of Professional Well Log Analysts, Paper TT, 15 p.

Katahara, K. W., 1995, Gamma Ray Log Response in Shaly Sands: The Log Analyst, v. 36, no. 4, p. 50–55.

Kennedy, W. D., 1995, Induction Log Forward Modeling: A Rigorous and Systematic Approach to Model Construction, *in* Transactions of the SPWLA 36th Annual Logging Symposium: Society of Professional Well Log Analysts, Paper G, 12 p.

Kennedy, W. D., D. C. Herrick, and T. Yao, 2001, Calculating Water Saturation in Electrically Anisotropic Media: Petrophysics, v. 42, no. 2, p. 118–136.

Kennedy, W. D., and D. C. Herrick, 2004, Conductivity Anisotropy in Shale-free Sandstone: Petrophysics, v. 45, no. 1, p. 38–58.

Klein, J. D., 1993, Induction Log Anisotropy Corrections: The Log Analyst, v. 34, no. 2, p. 18–27.

Klein, J. D, P. R. Martin, and D. F. Allen, 1997, The Petrophysics of Electrically Anisotropic Reservoirs: The Log Analyst, v. 38, no. 3, p. 25–36.

Looyestijn, W. J., 1982, Deconvolution of Petrophysical Logs: Applications and Limitations, *in* Transactions of the SPWLA 23rd Annual Logging Symposium: Society of Professional Well Log Analysts, Paper W, 20 p.

Luthi, S. M., 1990, Sedimentary Structures of Clastic Rocks Identified from Electrical Borehole Images, *in* A. Hurst, M. A. Lovell, and A. C. Morton, eds., Geological Applications of Wireline Logs: London, Geological Society Special Publication no. 48, p. 3–10.

McNaboe, G. J., 1991, Comparison of Formation Microscanner Images to Cores in Sandstone Reservoirs, Saudi Arabia: Society of Petroleum Engineers Middle East Oil Show Transactions, Bahrain, SPE Paper no. 21436, 7 p.

Mendelson, K. S., and M. H. Cohen, 1982, The Effect of Grain Anisotropy on the Electrical Properties of Sedimentary Rocks: Geophysics, v. 47, p. 257.

Meyer, W. H., 1987, Deconvolution of Invaded-bed Induction Problems *in* Transactions of the SPWLA 28th Annual Logging Symposium: Society of Professional Well Log Analysts, Paper P, 18 p.

Miller, M. N., Z. Paltiel, M. E. Gillen, J. Granot, and J. C. Bouton, 1990, Spin Echo Magnetic Resonance Logging: Porosity and Free Fluid Index Determination: Society of Petroleum Engineers annual Technical Conference and Exhibition, September 25–28, New Orleans, Louisiana, U.S.A., SPE Paper no. 20561.

Moran, J. H., and S. Gianzero, 1979, Effects of Formation Anisotropy of Resistivity Logging Measurements: Geophysics, v. 44, no. 7, p. 1266–1286.

Moran, J. H., 1982, Induction Logging—Geometrical Factors with Skin Effect: The Log Analyst, v. 23, no. 6, p. 4–10.

Morland, A. M. 1984, Special Log Processing for Thin Zones Using Geological Impulse Response Functions, with Particular Application to Total and Spectral Gamma Ray Logs: The Log Analyst, v. 25, no. 6, p. 23–35.

Mullarkey, J. C., and E. D. Dolly, 1997, Causes of Low Contrast, Low Resistivity in Hydrocarbon Productive Reservoirs of the Rocky Mountain and Mid-continent Regions: 1997 AAPG Annual Convention, April 6–9, 1997, Dallas, p. A86.

Passey, Q. R., K. E. Dahlberg, K. B. Sullivan, H. Yin, Y. H. Xiao, A. G. Guzman-Garcia, and R. A. Brackett, 2004, A Systematic Approach to Evaluate Hydrocarbons in Thinly Bedded Reservoirs, *in* Transactions of the SPWLA 45th Annual Logging Symposium: Society of Professional Well Log Analysts, Paper NNN, 12 p.

Pol, J. C., and S. R. Fekete, 1993, Depositional Environment and Reservoir/Seal Characteristics of the Miocene Potter Sand, Northern Midway-Sunset Field, Kern County, California; AAPG Pacific Section and Los Angeles Basin Geological Society Meeting, Abstract Only: AAPG Bulletin, v. 77, no. 4, p. 713.

Poupon, A., M. E. Loy, and M. P. Tixier, 1954, A Contribution to Electrical Log Interpretation in Shaly Sand: Gulf Coast Association of Geological Societies Transactions, p. 3.

Prammer, M. G., 1994: NMR Pore Size Distribution and Permeability at the Well Site: Society of Petroleum Engineers Annual Technical Conference and Exhibition, September 25–28, New Orleans, Louisiana, U.S.A., SPE Paper no. 28368.

Ramamoorthy, R., C. Flaum, and C. Coll, 1995, Geologically Consistent Resolution Enhancement of Petrophysical Analysis Using Image Log Data: Society of Petroleum Engineers Annual Technical Conference and Exhibition Transactions, Dallas, Texas, U.S.A., SPE Paper no. 30607.

Ruhovets, V., 1990, Log Analysis Technique for Evaluating Laminated Reservoirs in the Gulf Coast Area: The Log Analyst, v. 3, no. 5, p. 294–303.

Serra, O., 1989, Formation Microscanner Image Interpretation: Schlumberger Educational Services, SMP-7028, 117 p.

Serra, O., and M. Andreani, 1991, Thin Beds—A Guide to Interpretation of Thinly Layered Reservoirs: Schlumberger, 82 p.

Serra, O., I. Stowe, and D. Motet, 1993, True Integrated Interpretation *in* Transactions of the SPWLA 34th Annual Logging Symposium, Calgary: Society of Professional Well Log Analysts, Paper Z, 25 p.

Sovich, J. P., and B. Newberry, 1993, Quantitative Applications of Borehole Imaging, *in* Transactions of the SPWLA 34th Annual Logging Symposium, Calgary: Society of Professional Well Log Analysts, Paper FFF, 18 p.

Sneider, R. M., and J. T. Kulha, 1997, Low Resistivity, Low Contrast Pays: 1997 AAPG Annual Convention, April 9–12, 1997, Dallas, p. A109.

Sullivan, K. B., J. S. Campbell, and K. E. Dahlberg, 2003, Petrofacies: Enhancing the Deep-water Reservoir Characterization Effort in West Africa: Petrophysics, v. 44, no. 3, p. 177–189.

Tanguy, D., 1967, Methods and Apparatus for Investigating Earth Formations Featuring Simultaneous Focused Coil and Electrode System Measurements: U.S. Patent no. 3,329,889, http://www.uspto.gov/patft (accessed November 21, 2005).

Tarantola, A., 1987, Inverse Problem Theory: Amsterdam, Elsevier, 630 p.

Thadani, S. G., and G. A. Merchant, 1982, Deconvolution with Propagated Geometric Factors, *in* Transactions of the 57th Annual Technical Conference of the Society of Petroleum Engineers, September: SPE Paper no. 10986.

Trouiller, J. C., J. P. Delhomme, S. Carlin, and H. Anxionnaz, 1989, Thin-bed Reservoir Analysis from Borehole Electrical Images, *in* Transactions of the 64th Annual Technical Conference and Exhibition of the Society of Petroleum Engineers, San Antonio: SPE Paper no. 19578.

Van den Berg, F. G., W. J. Looyestijn, and R. K. J. Sandor, 1996, Sandwich: Log Evaluation in Laminated Shaly Sands, *in* Transactions of the SPWLA 37th Annual Logging Symposium: Society of Professional Well Log Analysts, Paper BB, 12 p.

Van Wagoner, J. C., R. M. Mitchum, K. M. Campion, and V. D. Rahmanian, 1990, Siliciclastic Sequence Stratigraphy in Well Logs, Cores, and Outcrops: AAPG Methods in Exploration Series, no. 7, 63 p.

Wang, H. M., L. C. Shen, and G. J. Zhang, 1998, Dual Laterolog Responses in 3-D Environments, *in* Transactions of the SPWLA 39th Annual Logging Symposium, May: Society of Professional Well Log Analysts, Paper X, 13 p.

Woodhouse, R., D. N. Greet, and C. R. Mohundro, 1984, Induction Log Vertical Resolution Improvement in Vertical and Deviated Wells Using a Practical Deconvolution Filter: Journal of Petroleum Technology, June, v. 36, no. 6, p. 993–1001.

Worthington, P. F., 1985, The Evolution of Shaly Sand Concepts in Reservoir Evaluation: The Log Analyst, v. 26, no. 1, p. 23–40.

Ye, S.-J., and P. Rabiller, 2000, A New Tool for Electro-facies Analysis: Multi-resolution Graph-based Clustering, *in* Transactions of the SPWLA 41st Annual Logging Symposium: Society of Professional Well Log Analysts, Paper PP, 14 p.

Yin, H., 2000, Limitations and Error Inherent in Resistivity Log Inverse Modeling for Formation Evaluation, *in* SEG Annual Meeting Transactions: Society of Exploration Geophysicists, p. 1798–1801.

Yin, H., K. E. Dahlberg, and R. A. Brackett, 2002, A Case Study of a Strange Induction Log: Modeling Array-induction Log Response to Sand Injection *in* Transactions of the SPWLA 43rd Annual Logging Symposium, June: Society of Professional Well Log Analysts, Paper F, 11 p.

Yin, H., and H. Wang, 2002, Method for 2-D Inversion of Dual Laterolog Measurements: U.S. Patent no. 6,430,509, http://www.uspto.gov/patft (accessed November 21, 2005).

Yu, L., O. N. Fanini, B. F. Kreigshauser, J. M. V. Koelman, and J. van Popta 2001, Enhanced Evaluation of Low-resistivity Reservoirs Using Multi-component Induction Log Data: Petrophysics, v. 42, no. 6, p. 611–623.

Zhdanov, M., D. Kennedy, and E. Peksen, 2001, Foundations of Tensor Induction Well-logging: Petrophysics, v. 42, no. 6, p. 588–610.

Index